8/6/04

PROTEIN CRYSTALLIZATION

IUL Biotechnology Series

PROTEIN CRYSTALLIZATION

TECHNIQUES, STRATEGIES, AND TIPS

A Laboratory Manual

edited by

TERESE M. BERGFORS

INTERNATIONAL UNIVERSITY LINE
La Jolla, California

Library of Congress Cataloging-in-Publication Data

Protein Crystallization: Techniques, Strategies, and Tips
    A Laboratory Manual / edited by Terese M. Bergfors
    (IUL Biotechnology Series)
    p. cm.
    Includes bibliographical references and index.
    1. Proteins.   2. Crystallization.   I. Bergfors, T.M.   II. Series.

QP552.P76   1999
572.633'5—dc21   98-075232
ISBN   0-9636817-5-3   AACR2

© International University Line, 1999-2001
Post Office Box 2525,
La Jolla, CA 92038-2525, USA

Library of Congress Catalog Card Number 98-075232

Printed in the United States of America

10 9 8 7 6 5 4 3 2

ISBN   0-9636817-5-3 $69.95   Hardcover

para Carmen Frida Julianna

Preface

Obtaining crystals is currently the bottleneck in protein structure determination by X-ray crystallography. The intent of this book is to collect
the most current methods for crystallizing proteins and present them as
lucid, easy-to-follow laboratory protocols. The accumulated knowledge
on practical aspects of protein crystallization is spread out in many different sources or in the form of local lab lore. The value of a laboratory
manual is that it organizes the practical portion of this knowledge.

This book began from my own experiences in crystallizing proteins
since 1984, and from teaching my course "Practical Protein
Crystallization". The contributing authors of the book have included
laboratory exercises where appropriate with their chapters so that the
book can be used in such courses or to test the techniques before beginning on expensive, hard-won proteins.

The first part of this book introduces the beginner to the basic techniques, materials, and parameters that affect crystallization. The more
experienced student can turn directly to the chapters on dynamic light
scattering and strategy. Dynamic light scattering is rapidly becoming an
increasingly important diagnostic tool and an introductory, step-by-step
guide is presented here. The protein, the star in this show, is sometimes
purified by the crystallographer and sometimes by biochemists or

molecular biologists. The protein purifier may not always know what handling considerations are important for a protein intended for crystallization experiments. Therefore I have included a chapter, based on questions from protein chemist collaborators through the years, to help identify the concerns in protein purification, which are unique for crystallization work.

Chapters 8 to 12 deal with crystallization strategy. It is essential to have a "map" or plan for searching the complicated multi-parametric space of crystallization conditions. If that plan does not work then change it, but by all means, begin with a plan of some sort. I have intentionally asked leading experts with conflicting approaches to present them. Thus, the contradictions that the reader finds among these chapters are not editorial oversights but a reflection of the current state of affairs in the screening problem.

Interpretation of results is probably the area in which beginners have the greatest difficulty. To my mind it is this ability to recognize the difference between a bad precipitate and a promising one, in the absence of any other leads, that constitutes the proverbial green thumb in crystallization. The pictorial guide in Chapter 13 helps you develop your own green thumb for growing protein crystals.

The second part of the book takes up crystallization for cryo-crystallography, seeding, the use of oils, and crystallization of membrane proteins. As soon as the beginner has mastered the basic techniques, the contents of this part are essential for completing the "crystallization toolbox".

The A-Z section was the most fun part of this book to compile. Since I learned some new things myself in editing this section, I daresay it contains something of value for everyone, regardless of their level of experience.

Some contents of the chapters overlap with material in other chapters. These repetitions have been permitted intentionally, with the reader in mind who will be dipping into the book, rather than reading it from cover to cover.

The editor (that's me) welcomes the readers to share their comments and experiences with the exercises and methods of this book by e-mail. (terese@xray.bmc.uu.se).

Acknowledgements

I am indebted to the following companies and organizations for permission to use their figures or tables in this book: Hampton Research, Chemicon International Inc., Millipore Intertech, Emerald BioStructures, Molecular Dimensions, International Union of Crystallography (IUCr), Current Biology Ltd., and Pierce Chemical Co.

I thank Drs. Matti Nikkola, Gerard Kleywegt and Professor Bror Strandberg for their critical reading of huge portions of this manuscript. My boss, Professor Alwyn Jones, allowed me to put aside the other interests of the lab for months so that I could work uninterruptedly on this book. Many colleagues in Uppsala contributed to this work and I thank Sherry Mowbray, Erling Wikman, Alex Cameron, Jill Sigrell, Mats Sandgren, Christina Divne, Inger Andersson, and Janos Hajdu for assistance and ideas. Stefan Knight, Brent Segelke, and Madeleine Riès-Kautt helped me with valuable discussions on crystallization strategy.

The authors of the chapters generously took time from their own projects in order to create this book and it has been a privilege to work with them. Dr. Igor Tsigelny, at the publishers, International University Line, initiated the idea for this manual and saw it to its completion. I thank him for his support of this project. Stefan Odestedt and Carl Bergfors made this book possible by contributing many hours of baby-sitting and I am also indebted to S.O. for his help with the table of contents. Finally, I extend my heartfelt gratitude to my daughter, Carmen for her forbearance during my preoccupation with this book.

Uppsala   Terese M. Bergfors
30 June 1998

Contents

Contributors

Jeff Abramson
Department of Biochemistry, Uppsala University
Biomedical Center Box 576, SE-751 23 Uppsala, Sweden.
jeff@xray.bmc.uu.se

Terese Bergfors
Department of Molecular Biology, Uppsala University
Biomedical Center Box 590, SE-751 24 Uppsala, Sweden.
terese@xray.bmc.uu.se

Alex Cameron
Department of Molecular Biology, Uppsala University
Biomedical Center Box 590, SE-751 24 Uppsala, Sweden.
alex@xray.bmc.uu.se

Naomi Chayen
Biophysics Section, Blackett Laboratory, Imperial College
London SW7 2BZ, UK.
n.chayen@ic.ac.uk

Kerstin Fridborg
Department of Molecular Biology, Uppsala University
Biomedical Center Box 590, SE-751 24 Uppsala, Sweden.
kerstin@xray.bmc.uu.se

Elspeth Garman
Laboratory of Molecular Biophysics, Department of Biochemistry,
University of Oxford
Oxford OX1 3QU, UK.
elspeth@biop.ox.ac.uk

Lesley Lloyd Haire
Division of Protein Structure, National Institute for Medical Research
The Ridgeway, Mill Hill, London NW7 1AA, UK.
l-haire@anika.nimr.mrc.ac.uk

So Iwata
Department of Biochemistry, Uppsala University
Biomedical Center Box 576, SE-751 23 Uppsala, Sweden.
iwata@xray.bmc.uu.se

Gerard Kleywegt
Department of Molecular Biology, Uppsala University
Biomedical Center Box 590, SE-751 24 Uppsala, Sweden.
gerard@xray.bmc.uu.se

Alexander McPherson
Department of Molecular Biology and Biochemistry,
University of California at Irvine
Irvine, California 92697-3900, USA.
amcphers@uci.edu

Sherry Mowbray
Department of Molecular Biology, Swedish Agricultural University
Biomedical Center Box 590, SE-751 24 Uppsala, Sweden.
mowbray@xray.bmc.uu.se

Madeleine Riès-Kautt
Laboratoire d'Enzymologie et de Biologie Structurales
CNRS Bat 34, F-91198 Gif-sur-Yvette, France.
ries@lebs.cnrs-gif.fr

Elisabeth Sauer-Eriksson
Umeå Center for Molecular Pathogenesis,
Umeå University
SE-901 87 Umeå, Sweden.
liz@ucmp.umu.se

Enrico Stura
Département d'Ingéniere et d'Etudes des Protéines
CEA, Saclay, F-91191 Gif-sur-Yvette, France.
stura@balthazar.saclay.cea.fr

Torsten Unge
Department of Molecular Biology, Uppsala University
Biomedical Center Box 590, SE-751 24 Uppsala, Sweden.
torsten@xray.bmc.uu.se

Johan Zeelen
Department of Structural Biology, Max Planck Institute of Biophysics
Heinrich-Hoffmann Straße 7, D-60528 Frankfurt am Main, Germany.
zeelen@biophys.mpg.de

1

A Bit of Advice on Crystallizing Proteins

Alexander McPherson

1

A Bit of Advice on Crystallizing Proteins

Alexander McPherson

University of California, Irvine, USA

The first protein crystals, of hemoglobin, were grown over 150 years ago and were immediately recognized as an exceptional and valuable discovery. Simply their appearance signified purity, uniqueness and a link between the living and the inanimate world from which life emerged (though evolution was not yet suspected).

Since then protein crystals have evolved from objects of wonder and speculation, demonstrations of molecular purity, to essential intermediates in the discovery of macromolecular structure. During the course of that evolution methods of crystallization developed as well—from empirical, trial and error approaches to the more rational, directed, and better physically characterized concepts of the past several years. Today it is fair to argue that macromolecular crystallization has become a true science, whose products affect a wide variety of fields from biotechnology, to molecular biology and medicine.

The contents of this book will advance that science even further. The practice of any area of research ultimately reduces to a person's hands and mind focused on simple objectives organized according to a coherent strategy. The chapters found here are designed to describe and elaborate, in a systematic manner, the means, tools, reagents, and procedures necessary to achieve these otherwise simple ends. The ultimate goal is crystals—protein, nucleic acid, or virus crystals.

When making a macromolecule crystallize, what, in a physical sense, is one doing? There are two phases to crystallization, nucleation and growth. In the first, molecules must overcome an energy barrier to form a periodically ordered aggregate of critical size that can "survive" in a thermodynamic sense. The second, growth, is achieved by making the solid state more attractive to individual molecules than the free, solution state. To promote either stage, a condition of supersaturation must be created in the "mother liquor" or crystallization medium.

Supersaturation is a non-equilibrium state characterized by a tendency to form a solid phase, and to increase the proportion of material in that phase until a balance is restored, i.e., the saturation or equilibrium concentration is reached. This is accomplished in the laboratory by altering the characteristics of the mother liquor in a way that discomforts the macromolecules, disrupts the solvent, and, in a manner of speaking, induces an amicable separation.

The major problem with macromolecular systems, unusual for more conventional molecules, is that there are usually two, and frequently more, solid states that can form in the system. Not only may crystals develop, but more often, precipitates, oils, or gels. The goal of the investigator is to persuade molecules, in a chemical and physical sense, that the crystalline state is their most attractive option.

If success is achieved, the researcher has produced conditions that convince the molecules that they are better off, happier if you will, in the crystal than in solution. They are, the physical chemist would say, at a lower energy state in the crystalline lattice than when free in solution: They are more secure in association with colleagues than aimlessly wandering in space.

It is wise to bear in mind a few clear principles when trying to crystallize a macromolecule; to remember precisely what you are trying to achieve. Here is a set of practical and straightforward objectives, of value no matter what the molecules.

1. *Homogeneity*—Perhaps the single most important property of the system under study is its purity. Crystallization presupposes that identical units are available for incorporation into a periodic lattice. While cultural diversity may be a desirable characteristic in human society, crystals abhor foreigners. A uniform population of conformist molecules is essential. Purify, purify, purify, and then purify more.

2. *Solubility*—Before a molecule can be crystallized, it must be solubilized. In practice this means the creation of a monodisperse solution free of aggregates, non-specific oligomers, and molecular clusters. Often, especially with membrane proteins, it is a persistent problem. Even with more well behaved proteins, at the

concentrations required for crystal nucleation polydispersity is a constant fellow traveler.

3. *Stability*—No homogeneous molecular population can remain so if its members alter their form, folding, or association state independent of one another. Hence it is important that macromolecules in solution not be allowed to denature, form oligomers, or undergo significant conformational change. Take whatever measures are necessary to keep the macromolecules as stable and immutable as possible.

4. *Supersaturation*—All crystals grow from a system displaced from equilibrium so that restoration requires formation of the solid state. Thus the crystal grower's first task is to find ways to alter properties of the macromolecule solution, such as pH or temperature, to create the supersaturated state.

5. *Association*—In forming crystals, molecules organize by orderly self-association to produce a periodically repeating, three-dimensional array. Thus it is necessary to facilitate positive molecular interactions while avoiding so far as possible the formation of precipitate, non-specific aggregation, or phase separation.

6. *Nucleation*—The number, the ultimate sizes, and possibly the quality of crystals depend on the mechanisms and rates of nuclei formation. The crystallographer must seek to induce and encourage limited nucleation by adjustment of the physical and chemical properties of the system.

7. *Variety*—Macromolecules may crystallize under a wide spectrum of conditions and form numerous polymorphs with a diversity of physical and diffraction properties. Thus the crystal grower must not cease his/her efforts with the first crystal, but should explore as many opportunities for crystallization as possible through additional biochemical, chemical, and physical parameters.

8. *Control*—The ultimate value of any crystals is inevitably dependent on their quality and sizes. Perturbations and disruptions of the mother liquor are, in general, deleterious. The crystallographer must endeavor to maintain the system at an optimal state, without fluctuations or shock, until the crystals have matured.

9. *Impurities*—Impurities arise from a host of sources: macromolecules from the preparation, conventional molecules from reagents, etc. Impurities may contribute to a failure to nucleate and are frequently the crucial obstacle to X-ray diffraction quality crystals. The crystal grower must do his/her best to discourage the presence of impurities in the mother liquor and the incorporation of impurities and foreign materials into the lattice.

10. *Preservation*—Some, indeed most, macromolecular crystals degrade, lose diffraction quality, or develop other sensitivities upon aging that depreciate their value. Once grown, crystals can be stabilized and preserved by temperature change, addition of more precipitating agent, or other suitable alterations in the mother liquor. It is important that they be protected from shock and disruption.

In addition to the principles elaborated above, some further advice to the novice crystal grower is in order. First, be careful in relying too much on commercial or commonly used crystallization screens, and too little on your own native intelligence, particularly your intuitive senses. Frequently, screens fail to succeed with entirely crystallizable macromolecules. If screens fail, then systematic approaches may be essential, and increasingly the choices fall to the investigator. For example, dialysis against distilled water, used to crystallize many proteins, is ignored by the screening matrices. Put your faith in common sense.

Keep an open mind, search the literature for ideas, and try different methods, other approaches, and novel reagents and additives. Often only small changes in composition or procedure, for example, changing from hanging drops to sitting drops, may be sufficient to achieve success. Don't exclude the ideas of others; don't develop favorites in terms of techniques that exclude the applications of others. As the great playwright Tennessee Williams said, "Make voyages, attempt them. That's all there is."

Finally, patience and perseverance are among the most essential components of any crystallization effort. If you're serious about achieving success, don't concede defeat after a few weeks, or even a few months. Try other preparations, purify further, attempt new methods. There is always something more. The ends will justify the cost.

2

Crystallization Methods

Torsten Unge

2

Crystallization Methods

Torsten Unge

Uppsala University, Uppsala, Sweden

This chapter will present the techniques and equipment needed to set up crystallization trials. Suppliers for materials mentioned here are listed in Appendix A2.

2.1. The Typical Vapor Diffusion Experiment

The most frequently used crystallization method is the vapor diffusion technique. In this technique a small droplet of typically 2-10 µl of the protein is mixed with an equal or similar volume of the crystallizing solution (usually buffer, salt, and precipitant) and placed on a siliconized glass cover slip. The cover slip is inverted and sealed into place over a reservoir containing 500-1000 µl of the crystallization solution. The difference in concentration between the drop and the reservoir drives the system toward equilibrium by diffusion through the vapor phase. Equilibration against a salt solution is faster (2-4 days for a 10 µl drop) than against polyethylene glycol solutions which can take up to 25 days. In the perfectly designed experiment the protein becomes supersaturated and crystals start to form when the drop and reservoir are at or close to equilibrium.[1]

Vapor diffusion drops can be prepared in several ways described below.

2.1.1 Hanging Drops

The volume of the hanging drop can be from a few μl up to 20 μl. For larger volumes, sitting or sandwich drops are used. The shape of the droplet is important because it influences the number of nucleation sites and consequently crystal size. If the solutions do not contain detergents or organic solvents, well-structured, spherical droplets can be readily-formed, by hanging from a siliconized glass cover slip.

2.1.2 Sitting Drops

Sitting drops can be used as an alternative to hanging drops in most cases. For large volumes (>20 μl) or in the presence of additives which lower the surface tension of the droplets (e.g., detergents and organic solvents), sitting drops are better. Sitting drops can be made on cover slips, in plastic microbridges[2] or in glass sitting rods.[3]

Under some conditions, and especially if the experiment is continuing over a long period, the surface of the droplet can become covered with a film of denatured protein. In hanging drops, it is not uncommon for the crystals to be firmly attached to the film. A solution to this problem can be to use sitting or sandwich drops.

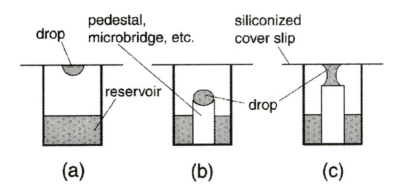

Figure 2.1. Three types of vapor diffusion setups:
(a) Hanging drop; (b) Sitting drop; (c) Sandwich drop.

2.1.3 Sandwich Drops

Drops can also be made sandwiched between two cover slips. The Q Plate (Hampton Research) offers this option. This technique is rarely used, but provides a method of reducing the area of the drop that is exposed to the air. Consequently the rate of diffusion is slower than in hanging and sitting drops.[4]

2.1.4 Reverse Vapor Diffusion

In the typical vapor diffusion experiment, a droplet with a lower concentration of the crystallization components is equilibrated against a reservoir solution of higher concentration. The drop will therefore become smaller as it equilibrates. It is also possible to arrange the opposite situation, i.e., equilibrate the drop against a reservoir of lower concentration or water.[5] This can be a way of crystallizing a protein that is less soluble at low ionic strength. The starting protein concentration must be quite high (at least double what is typically used) because the protein gets diluted as the droplet equilibrates against the reservoir. The droplets can become quite large. To avoid this, it is more common to use the dialysis method when crystallizing at low ionic strength.

2.1.5 pH Gradient Vapor Diffusion

In the regular vapor diffusion experiment only non-volatile buffer components should be used so that the pH remains the same throughout the experiment. Sometimes, however, one might like to perform the experiment by increasing or decreasing the pH. This can conveniently be done using acetate, ammonia, or carbonate buffers. Titrate the crystallization buffer one or two pH units below the starting pH by the addition of acetic acid and use this as the reservoir solution. As the acetic acid evaporates through the vapor phase, the pH in the drop will lower. To increase the pH, use ammonia or sodium carbonate in the reservoir.[6]

2.1.6. Practical Tips for Vapor Diffusion

2.1.6.1. Temperature Equilibration of the Materials

The vapor diffusion technique is sensitive to temperature variations. Thus the experiment must be arranged such that temperature differences between the cover slips and the reservoir solution are avoided. The Q Plates I and II (Hampton Research) are the best-designed crystallization plates in this respect. It is important that all solutions, plates, and cover slips be equilibrated to the crystallization temperature prior to the setup of the experiment.

2.1.6.2. Choice of Plate and Sealing

A wide assortment of crystallization plates is available. Most are 24-well tissue culture plates that will hold 1 ml in the reservoirs. Crystal Clear Strips (Hampton Research) is an exception: It has 96 wells and uses 100 µl in the reservoirs. All of these are made of polystyrene, which is a porous plastic. This means that there is ongoing evaporation of the system components and that, sooner or later, the drops and reservoirs dry out. An older method for vapor diffusion, without this problem, uses a glass spot plate (Corning cat. no. 7220) in a plastic sandwich box with a large volume (20 ml) reservoir.[7]

The setup must be properly sealed for diffusion to take place. Depending on the type of plate and choice of setup, this can be done with vacuum grease, high viscosity immersion oil, or clear sealing tape. It is difficult to make a perfect seal, so inspect each rim well under the microscope for bubbles or air vents in the grease or oil. Tape usually works better in this respect, making a good seal.

2.2. Other Methods

Although vapor diffusion is the most common, other methods can be used and these are presented below. Figure 2.2 shows the frequency of each respective method as reported in the BMCD (Biological Macromolecular Crystallization Database) up to 1995.[8]

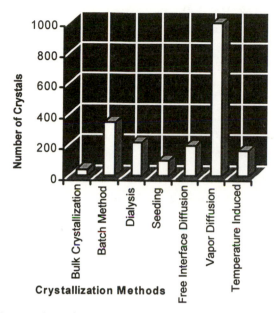

Figure 2.2. Number of crystals reported vs crystallization method.
Courtesy of Hampton Research, Current Biology Ltd .[9]

2.2.1. Dialysis

Dialysis offers a way of manipulating the ionic strength that is not possible with vapor diffusion. The dialysis can be set up in various ways, e.g., in dialysis bags or collodion thimbles (also called ultra thimbles), Zeppezauer cells, or buttons.[1] Dialysis buttons are commercially available for volumes from 5 μl to 350 μl. The protein is placed inside the button which is then covered by a dialysis membrane. Dialysis membranes are semi-permeable: They allow small molecular weight substances to diffuse in while preventing the protein from diffusing out. The protein's size determines the appropriate molecular weight cutoff for the dialysis membrane.

Dialysis is the most effective technique for crystallization by decreasing the ionic strength. If a protein is less soluble at low ionic strength, it is often possible to crystallize it simply by dialyzing it against a weak buffer or even water. No precipitant is necessary. Dialysis is also the method of choice for increasing the ionic strength while keeping the polyethylene glycol (PEG) concentration constant. This is because

PEGs of molecular weight 4000 and higher penetrate only extremely slowly through dialysis membranes, regardless of the dialysis membrane size cutoff.

Unlike in vapor diffusion, the protein concentration remains constant during the dialysis experiment.

2.2.2 Batch Techniques

In the batch technique the precipitant and the target molecule solution are simply mixed. Supersaturation is achieved directly rather than by diffusion. If the crystallization conditions are already known, this can be done preparatively, in large volumes (e.g., 350 µl or more) in a test tube or other reservoir. It can also be used as a rapid micro-scale screening under the microscope for the purpose of determining the solubility conditions. See Chapter 5, *Precipitants*.

2.2.3. Microbatch

Batch techniques can also be performed under oil. The oil prevents evaporation and extremely small drops (<2 µl) can be made, hence the term 'microbatch.' This method is described fully in Chapter 16, *Oils for Crystals*.

2.3. Summary

The choice of technique affects the final protein concentration, ionic strength, and precipitant concentration as well as the kinetics (rate) at which equilibration is achieved. The goal of all these different techniques is to put the protein into a state of supersaturation so that crystallization will be possible.

References and Further Reading

1. Ducruix A, Giegé R: *Crystallization of Nucleic Acids and Proteins. A practical approach.* Oxford: IRL Press Oxford University Press; 1992:82-90.
   Describes the vapor diffusion and dialysis methods.

2. Harlos K: **Micro-bridges for sitting-drop crystallizations.** *J Appl Cryst* 1992, **25**:536-538.
   Technical information on plastic microbridges for sitting drops.

3. Stura EA, Johnson DL, Inglese J, Smith JM, Benkovic SJ, Wilson IA: **Preliminary crystallographic investigations of glycinamide ribonucleotide phosphorylase.** *J Biol Chem* 1989, **264**:9703-9706.
   Describes glass sitting rods for sitting drops.

4. Fox K, Karplus A: **Crystallization of old yellow enzyme illustrates an effective strategy for increasing protein crystal size.** *J Mol Biol* 1993, **234**:502-507.
   Larger crystals were obtained from sandwich drops than hanging drops. The setup geometry affects the crystal size.

5. Jeruzalmi D, Steitz A: **Use of organic cosmotropic solutes to crystallize flexible proteins: application to T7 RNA polymerase and its complex with the inhibitor T7 lysozyme.** *J Mol Biol* 1997, **274**:748-756.
   Describes "reverse" vapor diffusion and crystallization by lowering the ionic strength.

6. McPherson A, Jurnak F, Singh GJP, Gill SS: **Preliminary X-ray diffraction analysis of crystals of *Bacillus thuringiensis* toxin, a cell membrane disrupting protein.** *J Mol Biol* 1987, **195**:755-757.
   This protein was crystallized by creating a pH gradient through the addition of 50 μl of 1 M acetic acid to the reservoir solution.

7. McPherson A: *Preparation and analysis of protein crystals.* New York: John Wiley and Sons; 1982:94.
   Alternative to polystyrene crystallization dishes.

8. Gilliland GL, Tung M, Blakeslee DM, Ladner, J: **The biological macromolecule crystallization database, version 3.0: new features, data, and the NASA archive for protein crystal growth data.** *Acta Cryst* 1994, **D50**:408-413.
   The BCMD provides useful statistics on crystallization of macromolecules.

9. McPherson A, Malkin AI, Kuznetsov AG: **The science of macromolecular crystallization.** *Structure* 1995, **3**:759-768.
   Review article.

Lab Experiment: Crystallization of Hen Egg White Lysozyme by Two Different Methods

Lysozyme is cheap and easy to crystallize. Many of the experiments in this book will be using lysozyme as the model protein. Below you will find two recipes for your first crystallization efforts.

Experiment 2.1. Vapor diffusion hanging drop technique

In the following experiment hen egg white lysozyme is crystallized using the hanging drop technique. The effect of different salt concentrations will be tested here. Other parameters such as pH, protein homogeneity, etc., are equally important.

Materials:
1. Q plate I or II
2. siliconized cover slips, 18 mm
3. sealing tape
4. lysozyme: chicken egg white, SERVA Fein Biochemica GmbH & Co KG, cat. no. 28262

Prepare the following solutions:
1. **Lysozyme:**
   Make 100 µl each of 15 mg/ml and 20 mg/ml lysozyme in 20 mM Na-acetate, pH 4.2
2. **Precipitant solutions:**
   A: NaCl, 5% (w/v) in 0.1 M Na-acetate, pH 4.2
   B: NaCl, 6% (w/v) in 0.1 M Na-acetate, pH 4.2
   C: NaCl, 7% (w/v) in 0.1 M Na-acetate, pH 4.2

Experiment:
1. With a pipette place 500 µl of precipitant solution A in the first reservoir. Carefully place one drop of 5 µl lysozyme (15 mg/ml) on a cover slip. Try to make the droplet as spherical as possible. Add 5 µl of the reservoir solution to the lysozyme droplet. It is important that the precipitant is added to the protein solution and

not in the reverse order. Exposure to high precipitant concentration can result in precipitation of the protein, which might destroy the experiment. Invert the cover slip and put it in place on the ledges in the Q plate. Prepare another two experiments with the same protein concentration and the precipitant solutions B and C.

2. Similarly prepare 3 new reservoirs with the solutions A, B, and C, and place 5 µl droplets of the 20 mg/ml lysozyme solution on the cover slips. Now seal the Q Plate with clear sealing tape.

Crystals will appear after 16 hours in the experiment with the highest salt and protein concentration. The quality of the crystals is highly dependent on the purity of the lysozyme. Purification of the lysozyme on a size exclusion column and concentration of the material with ammonium sulfate precipitation results in well-formed crystals with no packing disturbances. Even if you do not further purify the lysozyme, it should be filtered through a 0.22 µm filter immediately before setting up the crystallization trials.

Experiment 2.2. Crystallization by the batch method

1. Dissolve 25 mg of lysozyme in 350 µl of 0.2 M Na-acetate, pH 4.7, in a microcentrifuge tube.

2. Slowly add 350 µl of 10% NaCl, while vortexing at low speed. Centrifuge at 14,000 rpm for 5 minutes in tabletop centrifuge.

3. Transfer to suitable glass or clear plastic vials (e.g., scintillation vials or 2 ml Ellerman tubes) and seal them with parafilm. Crystals should appear after several hours or the next day.

3

Protein Samples

Terese Bergfors

3

Protein Samples

Terese Bergfors

Uppsala University, Uppsala, Sweden

It is always advantageous to purify the protein yourself because you learn a great deal about its behavior and stability in the process. If someone else purifies the protein for you, it can be hard for them to know which considerations are important when handling a macromolecule that is intended for crystallization trials. The following guidelines may be helpful.

3.1. Avoid Lyophilization

There are many examples of proteins which crystallize after lyophilization, lysozyme being one of them. Lyophilization is often used as a way to concentrate or store proteins, but it is not a gentle method and should be avoided if possible when the protein is intended for crystallization.

If the protein you receive is lyophilized, it is important to **dialyze** it against the target buffer. It is not sufficient to **dissolve** the protein in the buffer, for several reasons. The protein chemist may have lyophilized the protein in a non-volatile buffer. The buffer residue will then be a major contaminant when the protein is redissolved. It can also happen that the protein chemist changes the buffer in which the protein is lyophilized and does not inform you (because they are unaware of the significance for your experiments). This can result in a long series of irreproducible experiments while the cause of the problem is traced. Therefore, dialyze the lyophilized protein thoroughly, measure the concentration, and check the pH.

3.2. Ammonium Sulfate Precipitation

Be careful when using ammonium sulfate precipitation as a purification or concentration step.

As an early step in the purification, ammonium sulfate precipitation is usually perfectly acceptable, but it is not advisable as a final step when the protein is going to be crystallized. One type of problem is that it is virtually impossible to dialyze away, or remove on a short desalting column, all traces of the ammonium sulfate. Even minute amounts are sufficient to react with the components in the crystallization screen, giving rise to salt crystals or irreproducible results. Ask the protein chemist to replace this step with other concentration methods where possible.

3.3. Keep Purification Batches Separate Whenever Possible

Do not mix different purification batches in crystallization trials. Neither the growth conditions nor the purification are ever identical, so each new batch should be considered and screened separately.

3.4. Characterize the Protein Immediately upon Arrival

The more ways in which you can characterize your protein, the better, although restrictions on time and material will determine to what extent this is possible. Always run SDS-PAGE (sodium dodecyl sulfate polyacrylamide gel electrophoresis) and native PAGE or dynamic light scattering as soon as the protein arrives. If possible, complement with an IEF (iso-electric focusing) gel and mass spectroscopy. Store a small quantity of the protein from each batch at -70°C as archive material.

For the initial screening trials, the protein should be at least 90% pure when stained with Coomassie on an SDS gel. Always consider further purification of the protein 1) if the initial screen does not produce any promising results, or 2) to improve crystal quality when optimizing.

The initial gels are important for documentation of the homogeneity of the protein and possible batch variations. In addition, proteins degrade with time. If you notice that a certain batch of protein is no longer producing good crystals, run new gels (preferably native PAGE or IEF) on the protein. Comparison of these gels with the original ones will establish if the protein has deteriorated in some way. Similarly, gels can be run on failed crystallization drops or ones that suddenly produce crystals after a suspiciously long time.

3.5. Storage of Protein

Not all proteins tolerate freezing at -20°C. Most proteins can be kept at 4°C or -70°C, but the activity and stability must be checked for each new protein. Freezing and rethawing of the sample should be avoided, so store the protein in aliquots. Sometimes glycerol (10-50%) is added to the protein to help it tolerate freezing better. This can be a problem in itself because removing the glycerol by dialysis is slow and difficult.

Cells or bacteria tolerate freezing (-70°C) better than many purified proteins. It can be better to store the cells or other expression hosts, then thaw them and purify the target expression product (protein) freshly. Use of protease-deficient hosts can be expedient here.

My preference is to freeze the protein (without any glycerol) directly in liquid nitrogen and store it at -70°C. The method is as follows:

1. Take a 500 ml glass beaker and fill it with 100 ml liquid nitrogen. Do not use plastic (because the frozen pellets of protein tend to stick). Safety glasses should always be used when handling liquid nitrogen.

2. Add the concentrated protein solution drop-wise with a pasteur pipette directly into the liquid nitrogen from a height of 10-20 cm.

3. The frozen droplets should be about the size of green peas. Continue until all the solution is frozen. It is quite possible to

freeze some 25 ml of protein in 5-10 minutes. Pour or boil off the remaining nitrogen, then transfer the "peas" into a container (a 50 ml Falcon tube works fine) and place immediately at -70°C.

4. Aliquots ("peas") can be thawed as needed. Confirm the homogeneity and activity after long-term storage, regardless of the storage temperature.

As a general rule, it is better to store proteins concentrated than diluted. For example, the losses due to adsorption, etc., will be less if the protein is stored at 5 mg/ml than at 0.5 mg/ml. However, storing the protein *too* concentrated can lead to some of it precipitating.

3.6. Be Gentle

Handle the protein solution gently. Avoid foam. Do not vortex or shake it. If your protein stock solution is stored at a temperature different from the one at which crystallization trials will be set up, the best practice is to remove an aliquot and allow it to equilibrate to the target temperature. In practice, most people simply keep the protein stock on ice while setting up room temperature experiments and then return it to the cold room. In any event, avoid subjecting the protein stock to temperature variations unnecessarily.

3.7. Keep Good Records

Ask for a copy of the purification protocol. Is it the same for each batch? Collaborators often make changes without passing the information on, simply because they are unaware that it may have any significance. Make a note of everything you do to the protein.

3.8. Learn All about the Protein

Find out as much as you can about your protein. Some questions to ask are:

- Does it have free cysteines?
- Are there any known substrates/ligands/inhibitors?
- Is it sensitive to proteolysis?
- Does it bind metals or share homology with other metal-binding proteins which might indicate that it needs metals?
- Have similar proteins been crystallized?
- At what pH range is the protein stable/active?
- At what temperature is the protein stable/active?
- Is the protein glycosylated? phosphorylated? N-terminal methylated?
- Has sodium azide (or other bacteriocide) been added?

Read the literature on your protein and ask your collaborators questions. Important clues can emerge which will help obtain or optimize crystals.

4

Dynamic Light Scattering

Terese Bergfors

4

Dynamic Light Scattering

Terese Bergfors

Uppsala University, Uppsala, Sweden

One of the most useful diagnostic tools for crystallization is dynamic light scattering (DLS). Two factors are responsible for its emerging importance. The first is the improved theoretical understanding of aggregation in crystallization. The second is the advent of commercially available instruments that can perform DLS measurements with the quantities and solvent conditions that crystallizers typically use. This chapter will explain why DLS is so useful and how to perform and evaluate a DLS experiment.

4.1. What Does DLS Measure?

DLS measures the intensity of light scattered by molecules in solution, i.e., the translational diffusion coefficient. This is related to the hydrodynamic radius (Rh) of the molecule by the Stokes-Einstein equation. If aggregation occurs between molecules, it is recognizable by an increase in the Rh.

In other words, DLS measures the size distribution of the protein molecules in solution.

4.2. Why Use DLS to Measure Size Homogeneity?

The DLS profile of a protein is highly predictive of its crystallizability. Proteins with monomodal distributions (see Figure 4.1) have a

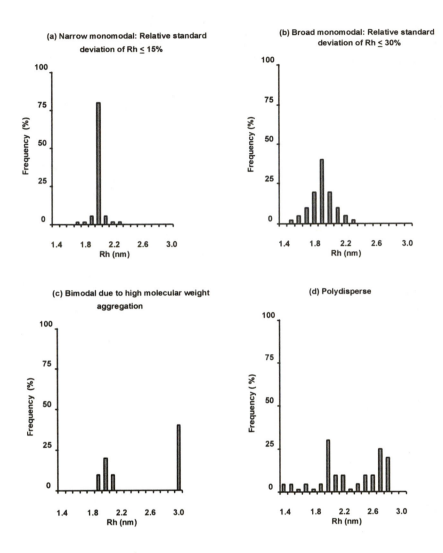

Figure 4.1. Different DLS profiles for a theoretical 15 kD protein with a monomeric Rh of 2.0 nm (20Å).

Monodisperse profiles can be classified as "narrow" or "broad" depending on the relative standard deviation of the Rh size distributions. Bimodality can be the result of several different phenomena (proteolysis, aggregation, monomer-dimer mixtures, etc.). Figure 4.1, c shows bimodality, which is the result of high molecular weight aggregates contaminating the 15 kD monomer.

high probability (70-80%) of producing some kind of crystals ([1] and Bergfors, unpublished results, for 66 and 36 different proteins, respectively). Conversely, proteins with non-specific aggregates are less apt to crystallize and the crystals are of poorer quality.

Non-specific aggregation is extremely detrimental to the formation of crystals. In order for a crystal to grow it requires an initial ordered nucleus upon which to add additional molecules. What causes molecules to come together and arrange themselves in ordered aggregates (nuclei) rather than disordered ones (precipitate)? The process is not well understood but studies have shown that for aggregation to lead to crystals it must not occur prior to supersaturation.[2,3] It is therefore important that the protein itself be free of aggregates prior to inducing supersaturation with crystallization precipitants.

The most frequent causes of non-specific aggregation are: 1) suboptimal protein solvent conditions, and 2) contamination by heterologous proteins.

The great value of DLS is that the initial solvent conditions can be optimized quickly (minutes) and with very little protein (20 µl at a few mg/ml for newer instruments). The assay is non-invasive, i.e., the protein can be recovered. Information about the size homogeneity of a protein can be obtained from other methods such as native polyacrylamide gel electrophoresis (PAGE) or size exclusion chromatography (SEC), but not with the speed or sensitivity levels of DLS.

4.3. Performing and Evaluating DLS Measurements on Your Protein

Instrument: Light scattering measurements can be performed on any single or multi-angle light scattering instrument. Some of the commercial instruments are: the Dawn F (Wyatt Technology Corp.), the BI-200SM (Brookhaven Instruments Co.), and the dp-801 (Protein Solutions, Inc.). When choosing an instrument intended for crystallization studies, look for useful features like:

- a small sample volume (20 to 250 µl);
- temperature control (a good range is 4-40°C);

- pertinent software for fitting the autocorrelation function to monomodal and bimodal distributions;
- custom support experienced in macromolecular crystallization applications.

The protocol described below in this chapter refers to the dp-801 instrument, but it is generally applicable.

Part 1: Sample Preparation

1. Run a native PAGE.
If you have enough protein, perform native PAGE on your sample. Otherwise go directly to the DLS measurements. Your protein should be at least 90% pure for meaningful DLS measurements.

Tip: Filter your PAGE samples with the same kind of filter that you will be using on the DLS samples in order to compare the results more fairly.

Comments:
Single band purity on native PAGE is neither a sufficient nor necessary prerequisite to obtaining crystals but has always been regarded as a good starting point. In our hands (Bergfors, unpublished results), the DLS profile seems to be a stronger predictor of crystallizability than native PAGE. In a study of 36 proteins for which both DLS and native PAGE were run, 15 proteins stained as single bands on PAGE with Coomassie and 20 had monodisperse DLS profiles (Rh relative standard deviation $\leq$ 30%). The DLS profiles are defined in Figure 4.1. Of the proteins which would have been considered "pure" by PAGE standards, 61% produced crystals, whereas 81% of the DLS "pure" proteins did.

It is interesting to note that the PAGE results are not always in agreement with the DLS profiles. For example, four of the proteins with narrow monomodal DLS profiles contained (faint) multiple bands on native PAGE. Nevertheless, these four proteins produced crystals that diffracted to better than 3Å. We conclude that the PAGE results are most useful in conjunction with the DLS results, rather than on their own. See the section 4.4.2, *If Your Protein Is Not MONOMODAL.*

2. Prepare sample: concentration and volume.

Prepare 25-200 μl of protein sample in a buffer of at least 10 mM. The volume required will depend on how your instrument is equipped. To save protein, you probably want to begin DLS measurements at a low protein concentration (see point 8 below). The minimum concentration required for a good DLS signal will depend on the molecular weight of the protein; a table is provided in the user manual. For example, a protein of 14 kDaltons should have a concentration of 3 mg/ml; for a protein of 100 kD, 0.4 mg/ml is sufficient. Larger molecules scatter light better than smaller ones.

If the buffer contains high salt concentrations or additives like glycerol, the default value for the viscosity should be changed in the software to prevent over-estimations of the Rh.

3. Set the starting temperature.

Turn on the dp-801 and set the thermostat to the temperature of interest. Flush the tubing with water and follow the manufacturer's recommendations for obtaining a background count rate. Now flush the tubing with buffer—many samples will precipitate if they come into contact with residual water in the tubing.

Your protein should be on ice if you are going to begin measuring at 4°C. After sample filtration and injection (see below), the sample temperature will be displayed on the screen. Confirm that your sample temperature is stabilized before beginning measurements. Take a set of 10-20 measurements for each different temperature to find at which one your protein is most monodisperse.

Some proteins need to "recover" after shifting the temperature. We found one protein that initially displayed a polydisperse profile. After more than 30 minutes at the new temperature, the polydispersity disappeared (Mowbray, unpublished results). The effect was repeatable and more pronounced when going from a high temperature to a lower one than in the reverse direction. Undoubtedly this phenomenon is protein-specific but be aware that it can occur.

Importance of the temperature profile:
Crystallizers who would routinely set up crystallization trials at pH values 4 to 9 often limit themselves to screening temperature

at only two points, usually 4°C and 20°C. We have not found any proteins yet where the dispersity profile remained constant over the range +4° to +37°C. Temperature is the easiest parameter to optimize with DLS and can eliminate many unnecessary crystallization trials.

4. Mount in-line filter. What size and kind of filter to use?

The recommended filters are Whatman's 0.1 μm Anotop™, for their exactness in pore size cutoff and low protein binding. Made of alumina oxide, they are virtually inert, but some His-tagged proteins might stick. For these cases, pre-rinsing the filter with EDTA can help.

The instrument manufacturers insist on in-line filtration of the protein to remove dust particles. Some aggregated species may also be removed by the filters, but particles of the size blocked by a 0.1 μm (100 nm) filter would render a sample unfit for crystallization anyway. For purposes of comparison, a 0.010 μm (10 nm) filter cutoff is roughly equivalent to a molecular weight cutoff (MWCO) value of 1,000 kD. For a further discussion on filtration, see the *A-Z Tips* section of this book.

The Anotops™ are also available in 0.02 μm size, as well as in different formats. Do not begin with this size (don't use it at all if your protein is larger than 150 kD) because if the sample is extremely aggregated it can completely clog the filter with resultant loss of sample. If your sample shows some aggregation in the profile after filtration with the 0.1 μm filter, recover it and repeat again with the 0.02 μm filter. This may be sufficient to remove the aggregation.

5. Inject the sample and begin taking measurements.

Please refer to the manufacturer's user manual for details on achieving a steady photon count rate and displaying of the data on the screen. We usually collect 10-20 measurements for each sample condition. This takes approximately 10 minutes. The sample can be recovered for more DLS measurements or crystallization work.

Part 2: Further Testing of Protein Solvent Conditions with DLS

Depending on the sort of profile you have obtained on your protein (see Figure 4.1) you may want to vary other protein solvent conditions besides temperature.

Recover the protein and try the following suggestions but avoid conditions known to precipitate your sample. Precipitation will result in the loss of sample and necessitate washing the tubing inside the instrument.

6. **Add 0.1 M ammonium sulfate.**

   Check the effect of ionic strength. If your sample is in aqueous buffer without salt, add 0.1 M ammonium sulfate, or 1 M NaCl. If the profile is sensitive to ionic strength, it will be evident at these concentrations.

7. **Change the pH.**

   Recover the protein and change the pH. Alternatively, you can prepare protein samples at different pH values at the outset of the experiment. However, if the amount of your sample is too limited to permit this, it will be necessary to exchange the buffer in one and the same sample. How to do this is explained in Appendix A1, *Good-to-Have Gizmos*.

   pH is the single most important parameter for protein solubility. It is only because of technical ease that temperature and ionic strength have been examined first in this experiment. If protein availability permits, and especially if you have no prior information on the pH preferences of your protein, different pH values should be examined from the start.

8. **Increase protein concentration.**

   Begin your DLS experiments at the lowest recommended concentration in the manufacturer's table, because if your protein already contains non-specific aggregates at a low concentration, the problem is only aggravated by higher concentrations. However, if the protein is monodisperse at low concentrations, you may want to recheck the profile at higher protein concentrations, e.g., from 5 to 20 mg/ml. Increasing the protein concentration can affect monomer/dimer equilibriums and may improve the distribution ratio between the oligomers.

You may need to change the viscosity default in the software for the more concentrated samples.

9. **Data interpretation**

The DYNAMICS software provided with the dp-801 calculates an autocorrelation function to fit the measurements to either monomodal or bimodal distributions. If the size dispersity is so great that it cannot be resolved into either sort, then the sample is most likely polydisperse (see Figure 4.1). Several different software programs exist and may be useful in difficult cases of interpretation.

4.4. Recommendations for Handling Your Protein Based on Its DLS Profile

4.4.1. If Your Protein Is MONOMODAL

- **Great.** Set up crystallization trials.
- **If you do not get crystals:**
  A perfectly monodisperse DLS profile is not a guarantee of obtaining crystals because there can be other types of heterogeneity which may be unrelated to size dispersion but still prevent the formation of good molecular contacts within the crystal.

Look for other sources of protein heterogeneity like:
  a. mixture of apo and holo forms of the protein
  b. differences in glycosylation patterns
  c. conformational flexibility in the protein
  d. differently charged isoforms
  e. mixture of methylated and non-methylated forms
  f. differentially phosphorylated forms

Example: Some *Trichoderma reesei* cellulases have been notoriously difficult to crystallize, even though they have DLS profiles with 0% polydispersity. In order to obtain crystals it was necessary to (a) proteolytically remove a probably floppy domain **and** (b) deglycosylate the protein.

4.4.2. If Your Protein Is Not MONOMODAL

In addition to optimizing the protein solvent conditions as described in Part 1:

- **Look at your native gel.**
  Is your protein aggregating because of contamination by heterologous proteins? If yes, purify further. Mixtures of monomers/dimers or monomers/oligomers should be avoided. Are these oligomeric states functions of the solvent conditions or do you need another chromatographic step to separate them?

- **Try additives.**
  Some additives to try are glycerol, detergent, or 1M LiCl. It may not be possible to take DLS measurements of the protein after the addition of a detergent. The detergent micelles themselves can be larger than the protein. Check the DLS profile of the detergent alone first to see if this is the case.

- **Consider different constructs.**
  Remove C-terminus or N-terminus, truncate domains, remove (or move) the His-tag. For a good example, see reference.[4]

- **Set up crystallization trials anyway.**
  Sometimes highly impure, polydisperse proteins do produce crystals. The crystallization itself can be a purification step.

4.5. Summary

DLS is the best "quality control assay" currently available for your protein. Use it.

References and Further Reading

1. D'Arcy A: **Crystallizing proteins: a rational approach?** *Acta Cryst* 1994, **D50**:469-471.

   This DLS study of 66 proteins measured the profiles of the proteins alone (no precipitant) and correlated them to the subsequent crystallization results.

2. Mikol V, Hirsch E, Giegé R: **Diagnostic of precipitant for biomacromolecule crystallization by quasi-elastic light-scattering.** *J Mol Biol* 1990, **213**:187-195.

   This study examines aggregation of a protein (lysozyme) in the presence of precipitants.

3. Wilson WW: **Monitoring crystallization experiments using dynamic light scattering; assaying and monitoring protein crystallization in solution.** *Methods Comp Meth Enzymol* 1990, **1**:110-117.

   Another study on the aggregation of lysozyme in the presence of precipitants.

4. Ferré-D'Amaré AR, Burley, SK: **Use of dynamic light scattering to assess crystallizability of macromolecules and macromolecular assemblies.** *Structure* 1994, **2**:357-359.

   Good example of examining different constructs by DLS to find the most appropriate one for crystallization trials.

5

Precipitants

Terese Bergfors

5

Precipitants

Terese Bergfors

Uppsala University, Uppsala, Sweden

The function of the different precipitants in the crystallization drop is to alter the protein-solvent or protein-protein contacts so that the protein molecules precipitate out of solution, preferably as ordered crystals and not as disordered aggregates. This chapter presents some practical considerations in choosing a precipitant and a working concentration range. A list of precipitants, compiled from the literature, is given in Table 5.1. The precipitants can be grouped into four categories based on their mode of action: salts, organic solvents, polymers, and surfactants.

5.1. Which Precipitant to Use?

The choice of precipitant must be made empirically. Figure 5.1 lists the 10 most common precipitants reported in the Biological Macromolecule Crystallization Database (BMCD)[1] through 1995. Ammonium sulfate is by far the most used precipitant, followed by polyethylene glycols (PEGs) 4000-8000. Preliminary analysis of crystallization reports through 1998 indicates that PEGs 4000-8000 will soon overtake ammonium sulfate as the number one precipitant (Gilliland, G., personal communication).

The pI of the protein can also be used as a starting point for the choice of precipitant; see Chapter 10, *An Alternative to Sparse Matrix Screens.*

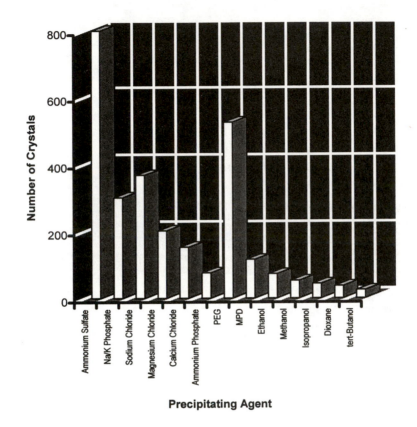

Figure 5.1. Number of crystals vs. precipitating agent. Courtesy of Hampton Research, Current Biology Ltd.[10]

5.2. Which Concentration Range to Use?

The BMCD statistics can serve as a guide to the right range of concentrations. These are presented in Figure 5.2 for three precipitants: ammonium sulfate, PEG 6000 and MPD (2-methyl-2,4-pentanediol). Thus, without any *a priori* information on your protein, a reasonable range of MPD concentration to begin testing is 40-60%, not 5-20%.

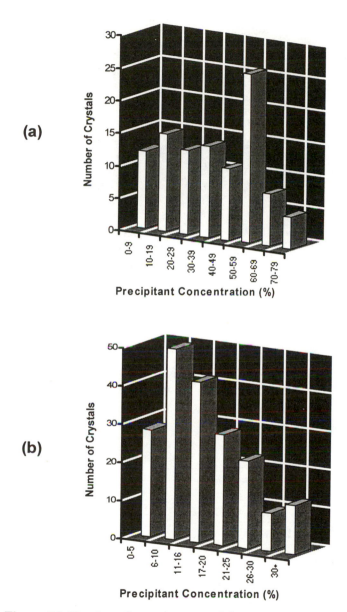

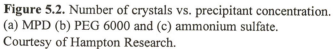

Figure 5.2. Number of crystals vs. precipitant concentration.
(a) MPD (b) PEG 6000 and (c) ammonium sulfate.
Courtesy of Hampton Research.

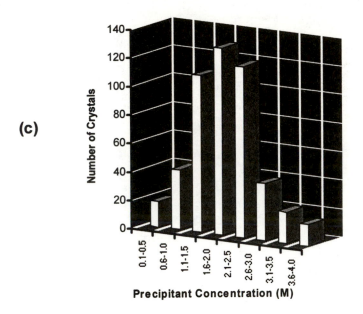

(c)

The precipitation point of your protein for any given precipitant can also be determined empirically as described below. This method is not extremely accurate, but it is quick and will give a rough idea of which concentration range to use.

5.3. Quick Protocol for Determining the Precipitation Point of a Protein

Materials needed:

1. A spot plate, microscope slide with depression well, or an ordinary cover slip can be used. The glassware should be silanized.

2. A 10 µl Hamilton syringe fitted with a repeating adaptor (PB600-1) to make 0.2 µl drops. (See Appendix A1: *Good-to-Have Gizmos*.)

3. A piece of black paper as a background, or a microscope.

Procedure:

1. Put the spot plate on a black background, or under the microscope.

2. Put a 5 µl drop of protein on the glass. The protein concentration should be relatively high (e.g., 20 mg/ml) and well buffered.

3. Add 0.2 µl of precipitant stock (e.g., 50% PEG, 3.5 M ammonium sulfate, or 80% MPD) with the repeating Hamilton syringe. Stir the drop well, with the point of the syringe.

4. Repeat, in 0.2 µl increments, until precipitation occurs. Do not proceed too rapidly; wait at least 30 seconds between injections. Cover the experiment to reduce evaporation. Record the number of injections.

5. Determine how many µl of precipitant have been added to the original 5 µl drop to calculate the percentage precipitant required to induce precipitation. This will give the concentration range around which to begin screening. Likewise, determine what the protein concentration is now, since it has become diluted.

This method works well with salts and MPD. PEG presents problems because of its viscosity but on the other hand, PEG has a wider window of critical concentration. The critical concentration range for a salt can be as small as 0.5%.

Examples: If you find that your protein precipitates with 20% PEG in this quick test, prepare a vapor diffusion screen with 15%, 20%, and 25% PEG in the reservoir. If your protein precipitates with 1.2 M ammonium sulfate, screen from 1.0 to 1.4 M. Repeat for different pH values and protein concentrations.

5.4. Grid Screens

Location of the precipitation point at a given pH, or for several differ-ent pH intervals, can be used as the basis for a grid design. A typical grid screen evaluates a single precipitant at 4-6 different concentrations vs. pH. Look for two contiguous drops where one remains clear but the next one precipitates: This is the boundary between nucleation and precipitation. The concentration range between these two drops should be more finely screened in the next round of crystallization trials.

Grid screens can be used for initial screening or optimization. References[2] and[3] contain good examples of grid screens. The following grid screens are commercially available from Hampton Research:

1. PEG 6000
2. MPD
3. Ammonium sulfate
4. NaCl
5. PEG 6000 + 1M LiCl
6. NaK phosphate

5.5. Comments

- Precipitants from the different categories can be used in combination with each other (i.e., as co-precipitants). Try salts with PEG; organic solvents, and salts or PEG; isopropanol with MPD, etc.

- The usefulness of any given precipitant is not necessarily reflected by its frequency of occurrence in the BMCD. It may well be underrepresented because it has not been tried as much, not because it is less effective. Don't be afraid to try new things.

- You don't always need a precipitant, for example if you use dialysis or vapor diffusion against water or low ionic strength buffer. Antibodies are almost always crystallized this way. This technique is described in Chapter 2, *Crystallization Methods* and in reference.[4]

Other examples of crystallization methods that do not use precipitants are found in Chapter 7, *Temperature* and in reference.[5]

References and Further Reading

1   Gilliland GL, Tung M, Blakeslee DM, Ladner, J: **The biological macro-molecule crystallization database, version 3.0: new features, data, and the NASA archive for protein crystal growth data.** *Acta Cryst* 1994, **D50**:408-413.

This database contains the crystallization conditions for 1465 macromolecules and is a valuable source of statistics.

2. Weber PC: **A protein crystallization strategy using automated grid searches on successively finer grids.** *Methods: A Comp to Meth Enzymol* 1990, **1**:31-37.

Describes grid screening.

3. Shaw Steward PD, Khimasia M: **Predispensed gradient matrices - a new rapid method of finding crystallization conditions.** *Acta Cryst* 1994, **D50**:441-442.

Describes grid screening.

4. Jeruzalmi D, Steitz TA: **Use of organic cosmotropic solutes to crystallize flexible proteins: application to T7 RNA polymerase and its complex with the inhibitor T7 lysozyme.** *J Mol Biol* 1997, **274**:748-756.

Describes the reduction of ionic strength by "reverse vapor diffusion."

5. Behlke J, Marg A, Paeschke M: **Protein crystallization with and without precipitants.** *J Appl Cryst* 1997, **30**:559-560.

Adrenoxin was crystallized in hanging drops without any precipitant by decreasing the vapor pressure with a vacuum pump.

6. McPherson A: **Current approaches to macromolecular crystallization.** *Eur J Biochem* 1990, **189**:1-23.

Review article with list of precipitants.

7. Riés-Kautt M, Ducruix A. **Inferences drawn from physico-chemical studies of crystallogenesis and precrystalline state.** *Meth Enzymol* 1997, **276**:23-59.

Use of the Hofmeister series and net charge of the protein to choose an appropriate precipitant.

8. Patel S, Cudney B, McPherson A: **Polymeric precipitants for the crystallization of macromolecules.** *Biochem Biophys Res Comm* 1995, **207**:819-828.

Novel polymeric precipitants were tested on 24 test proteins of which 14 crystallized.

9. Mustafa A, Derrick J, Tiddy G, Ford RC: **A novel approach for the crystallization of soluble proteins using non-ionic surfactants.** *Acta Cryst*
1998, **D54**:154-158.

Five test proteins were crystallized with these surfactants as precipitants in batch
and vapor diffusion techniques.

10. McPherson A, Malkin AI, Kuznetsov AG: **The science of macromolecular crystallization.** *Structure* 1995, **3**:759-768.

Review article.

A 5.1.

Table 5.1. Precipitants for protein crystallization

Salts[6,7]

        acetates

                ammonium, sodium

        bicarbonates

        cetyltrimethyl ammonium salts

                bromide, chloride

        chlorates

        chlorides

                ammonium, calcium, lithium, potassium, sodium

        chromates

        citrate

                ammonium, sodium

        formate

                sodium

        halides

                NaCl, NaBr, NaI, NaF

        maleate

        nitrates

                ammonium, potassium, sodium

        phosphates

                ammonium, potassium, sodium

        proprionate

Continued on next page

sulfates
>ammonium, calcium, cadmium, lithium, magnesium, sodium

sulfonates
>p-toluene, propane, benzoate, betaine
>
>sulfonates usually used as buffers:
>>MOPS, CAPS, MES

tartrates
>potassium, potassium-sodium (same as sodium-potassium; also called Rochelle salt)

thiocyanates
>ammonium, potassium, sodium

Organic solvents/small molecular weight polyalcohols[6]

>acetone
>acetonitrile
>butanol, tert-
>1,3-butyrolactone
>dimethyl sulfoxide (DMSO)
>dioxane
>ethanol
>2,5-hexanediol
>isopropanol
>methanol
>2-methyl-2,4-pentanediol (MPD)
>polyethyelene glycol (PEG) 400
>1,3-propanediol

Polymeric precipitants[6,8]

>carboxymethylcellulose
>>low, medium, high viscosity
>Jeffamine
>polyacrylic acid
>>2100, 5100
>polyamine
>polyethylene glycol (PEG)
>>1000-20,000 molecular weight

Continued on next page

PEG, methoxy
PEG, dimethyl ether
PEG, monomethyl ether
polypropylene glycol P400
polyvinyl alcohol 15,000
polyvinylpyrrolidone K15

Non-ionic surfactants (mild detergents)[9]

n-alkyl polyoxyethylenes

pentaoxyethylene mono-6-dodecyl ether ($C_6C_5EO_5$)

tetraoxyethyelene mono-n-octyl ether (C_8EO_4)

glycerol-1-nonaoxyethylene-2,3-diheptyl ether (DiC_7EO_9)

6

Buffers and pH

Terese Bergfors and Kerstin Fridborg

6

Buffers and pH

Terese Bergfors and Kerstin Fridborg

Uppsala University, Uppsala, Sweden

pH is the single most important determinant of protein solubility. Therefore it will be one of the first parameters to explore in crystallization trials. This chapter will cover some of the practical considerations for the choice of pH and buffer.

6.1. Buffers

6.1.1 Buffer Concentration for the Protein Solution

The protein must be dissolved in some buffer at the outset of crystallization trials. We usually use 10 mM HEPES, pH 7.0, or Tris, pH 8.0, if nothing known about the protein's stability contra-indicates this. It may be necessary to add some salt, e.g., 75-200 mM NaCl, to keep the protein in solution.

6.1.2 Buffer Concentration for the Drops

A 0.1 M concentration of buffer in the reservoir is typically used. However, it is important to note that some of the common salts used as precipitants contain volatile species which will affect the pH. In these cases (with ammonium sulfate, ammonium acetate, etc.), it will be necessary to adjust the pH of the buffer **after** the addition of the salt. Mixing ammonium sulfate with Tris buffer, pH 8.3, gives a solution with a pH of 6, not 8.3.[1]

6.1.3 Choice of Buffer

The pH itself is a major variable but the crystallization results are also affected by what kind of buffer is used. Figure 6.1 shows the difference in crystals obtained at the same pH (4.6) but with different species of buffers.

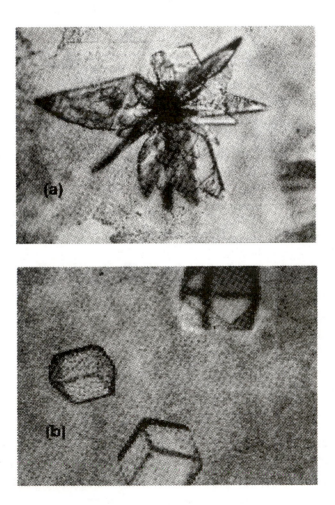

Figure 6.1. Effect of buffer choice on crystal quality.
All conditions were identical except for the choice of buffer, which was succinic acid in (a) and citrate in (b).

Another consideration in the choice of buffer is that inorganic buffers, although frequently used successfully in protein crystallization, often give rise to salt crystals. Phosphate buffers are particularly notorious in this respect. Citrate is a strong chelator of many metal ions. Therefore it is a poor choice if you are screening metal additives or making heavy atom derivatives (see Chapter 11, *Reverse Screening*, section 11.6.) Cacodylate, an arsenic compound, is often used in crystallizations, but it is poisonous and should be handled with appropriate caution.

All buffers can act as inhibitors of certain enzymes or otherwise bind to proteins. See reference[2] for lists of these. The buffer can appear as an unexpected electron density feature in the crystal, perhaps even in the active site. Therefore it is a good idea to keep careful records of every buffer your protein has been in during the purification.

Example: An unexpected density feature in the crystals of a complex of hCRABP II (human cellular retinoic acid binding protein II) turned out to be a phosphate moiety. Phosphate buffer was used in the initial chromatography step. In the next three chromatography steps, other buffers were used and the protein was thoroughly dialyzed against Tris buffer before crystallization trials. Still a phosphate showed up in the electron density. The purification records made it possible to interpret the unexplained density as phosphate, rather than, for instance, as sulfate.[3] See "buffers" in *A-Z Tips* for more examples.

Moral—there are two: Keep good purification records, and buffers matter.

The appendix of this chapter has four tables of buffers at different pH intervals versus their frequency of occurrence in the BMCD (Biological Macromolecule Crystallization Database).[4] These tables can be useful starting points for designing your own pH screen. If no successful crystallizations are reported for a given buffer, it may only mean that the buffer has never been tried and not that it is a bad choice.

6.2. pH

6.2.1 Initial Choice of pH Range for Screening

The strategy section of this book discusses several approaches to the question of which pH range to screen in crystallization trials. One can begin by determining the pI of the protein and screening the pH range appropriate for that pI (see Chapter 10). Another approach, based on the statistics from the BMCD, is to make a coarse screening in the pH range 4 to 9 at intervals of 1 or 1.5 pH units. The BMCD reports that proteins have been crystallized from pH 2 to l0, with 90% of them at pH 4 to 9.[4] See Figure 6.2.

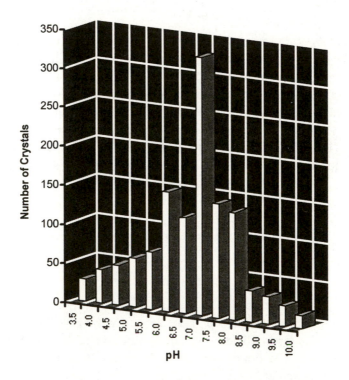

Figure 6.2. Number of crystals reported at various pH values. Courtesy of Hampton Research, Current Biology Ltd.[7]

6.2.2 Optimization of pH

After the first crystals are obtained, it may be possible to improve their quality by fine-tuning the pH. Concentrate your efforts at the extremes of the pH range where the crystals appear. For example, let us say you have tested integral pH values from 4 to 9 and crystals appeared from pH 5 to 8 but none at pH 4 or 9. Begin to examine the effects of pH in small increments (0.05 to 0.1 pH units) in the pH intervals 4-5 and 8-9. Remember that pH is logarithmic, not linear. For an excellent example of optimization by fine-tuning of the pH, see reference.[5]

References and Further Reading

1. Mikol V, Rodeau J-L, Giegé R: **Changes of pH during biomacromolecule crystallization by vapor diffusion using ammonium sulfate as the precipitant**. *J Appl Cryst* 1989, **22**:155-161.

    Ammonium sulfate causes pH changes in vapor diffusion drops due to a transfer of ammonia. It is important to be aware of this effect so that unintentional pH changes are not introduced. Alternatively, the effect can be exploited to induce crystallization by generation of a pH gradient.

2. Perrin B, Dempsey DD: *Buffers for pH and Metal Ion Control*. London: Chapman and Hall; 1987.

    An excellent reference book on buffers with recipes for making them. It also contains a useful chapter on which buffers are incompatible with which enzymes and metals.

3. Kleywegt G, Bergfors T, Senn H, Le Motte P, Gsell B, Shudo K, Jones TA: **Crystal structures of cellular retinoic acid binding proteins I and II in complex with all-trans-retinoic acid and a synthetic retinoid**. *Structure* 1994, **2**:1241-1258.

    Example where the phosphate buffer bound to the protein.

4. Gilliland GL, Tung M, Blakeslee DM, Ladner, J: **The biological macromolecule crystallization database, version 3.0: new features, data, and the NASA archive for protein crystal growth data**. *Acta Cryst* 1994, **D50**:408-413.

    This database contains the crystallization conditions for 1465 macromolecules and is a valuable source of statistics.

5. McPherson A. **Increasing the size of microcrystals by fine sampling of pH limits.** *J Appl Cryst* 1995, **28**:362-365.

    If you read only one article ever about pH and crystallization, read this one.

5. Gomori G: **Preparation of buffers for use in enzyme studies.** *Meth Enzymol* 1955; **1**:138-146.

    In this old book you can find help for making buffers of different pH values.

7. McPherson A, Malkin AI, Kuznetsov AG: **The science of macromolecular crystallization.** *Structure* 1995, **3**:759-768.

    Review Article.

A 6.1. Tables of Buffers versus Number of Successful Crystallizations in the BMCD.[4]

Table 6.1. The most commonly used buffers in crystallization experiments sorted by frequency of occurrence.

Buffer	pH range	Number of crystallizations
Tris-HCl	7.2-9.0	322
acetate	3.6-5.6	204
cacodylate	5.0-6.0	162
phosphate	6.9-8.0	122
HEPES	6.8-8.2	88
citrate	2.6-7.0	55
PIPES	6.1-7.5	31
Tris-maleate	7.0-8.6	14
Tris-acetate	not given	13
imidazole-HCl	6.2-7.8	12
MOPS	6.5-7.9	11
bis-Tris	5.8-7.2	10

Table 6.2. Buffers available from Hampton Research and their occurrence in the BMCD. (Note: The pH ranges given here are not always the same as given in the BMCD.)

pH-range	Buffer	Number of crystallizations
3.0-6.2	Sodium-citrate	86
3.6-5.6	Sodium-acetate	149
5.0-7.4	Sodium-cacodylate	112
5.2-7.1	MES	4
5.6-7.5	ADA	2
6.2-7.8	Imidazole-HCl	12
6.6-8.5	HEPES	92
7.0-9.0	Tris-HCl	322
7.4-9.3	Bicine	8

Table 6.3. Inorganic buffers[2,6] versus number of crystallizations.

pH range	Buffer	Number of crystallizations
1.0-2.2	HCl - KCl	
2.0-5.1	Tartaric acid - NaOH	
2.1-6.5	Citric acid - NaOH	
2.2-3.6	Glycine - HCl	4
2.2-3.1	KH-phtalate - HCl	
2.5-5.7	trans-Aconitic acid - NaOH	
2.6-4.8	Formic acid - NaOH	1
2.6-7.0	Citrate - phosphate	3
3.0-6.2	Citrate	55
3.2-4.0	K-phtalate - HCl	
3.2-7.6	3,3 Dimethylglutaric acid - NaOH	
3.4-5.2	Phenylacetic acid - Phenylacetate	
3.6-5.6	Sodium-acetate	149
3.8-6.0	Succinic acid - NaOH	1
4.0-10.7	Carbonate - Bicarbonate	1
4.1-5.9	KH-phtalate - NaOH	
4.3-5.7	Aconitate	
5.0-6.0	Succinate	10

Continued on next page

pH range	Buffer	Number of crystallizations
5.0-7.4	Cacodylate	172
5.2-6.0	Phtalate - NaOH	
5.2-6.0	NaH-maleate - NaOH	
5.2-8.6	Maleic acid - Tris - NaOH	2
5.7-8.0	Phosphate	202
6.2-6.8	Maleate	3
6.2-7.8	Imidazole - HCl	12
6.5-8.3	Trimethylpyridine - HCl	
6.8-8.6	N-Ethylmorpholine - HCl	
6.8-8.8	Triethanolamine - HCl	2
6.8-9.2	Barbital	
6.8-9.2	Sodium 5,5-diethylbarbiturate - HCl	1
6.9-9.0	Na-pyrophosphate - HCl	3
7.0-7.9	Bicine - NaOH	8
7.8-8.8	Ammediol	
7.8-9.7	2-Amino-2-methylpropane-1,3 diol - HCl	
7.8-9.9	Diethanolamine - HCl	
7.8-10.0	K-p-phenolsulphonate - NaOH	
8.0-9.1	Na-borate - HCl	
8.0-10.2	Boric acid - NaOH	
8.2-9.0	Ammonia/ammonium chloride	10
8.6-10.6	Glycine - NaOH	4
8.7-9.2	Boric Acid - borax	
9.2-10.8	Na-borate - NaOH	
9.2-10.8	Na_2CO_3 - $NaHCO_3$	1
9.3-10.1	Borax - NaOH	
9.6-11.0	$NaHCO_3$ - NaOH	
10.9-12.0	Na_2HPO_4 - NaOH	
12.0-13.0	NaOH - KCl	

Table 6.4. Good's biological buffers versus number of crystallizations.

Good's biological buffers were developed as a complement/replacement for inorganic buffers. Here are some advantages of the biological buffers:

- The pKa value lies mostly between 6 and 8, a pH range where most biochemical reactions take place.

- They are highly soluble in water, stable, and not affected by salt.
- For most of them pH is not sensitive to concentration or temperature. Tris is an exception.

Chemical supply catalogues, e.g,. Sigma Chemical Co., contain the list of biological buffers.

pH range	Buffer	Number of crystallizations
5.5-6.7	MES	4
5.8-7.2	BIS-TRIS	10
6.0-7.2	ADA	2
6.1-7.5	ACES	
6.1-7.5	PIPES	31
6.2-7.6	MOPSO	
6.3-9.5	BIS-TRIS PROPANE	7
6.4-7.8	BES	2
6.5-7.9	MOPS	11
6.8-8.2	TES	2
6.8-8.2	HEPES	88
6.9-8.3	MOBS	
7.0-8.2	DIPSO	
7.0-8.2	TAPSO	
7.1-8.5	HEPPSO	
7.2-9.0	TRIS	322
7.2-8.5	POPSO	
7.3-8.3	TEA	5
7.3-8.7	EPPS	
7.4-8.8	TRICINE	2
7.5-8.9	GLY-GLY	3
7.6-9.0	BICINE	8
7.6-9.0	HEPBS	
7.7-9.1	TAPS	1
7.8-9.7	AMPD	
8.2-9.6	TABS	
8.6-9.0	CHES	1
8.7-9.7	AMPSO	
8.9-9.3	CAPSO	
9.0-10.5	AMP	6
9.7-11.1	CAPS	2
10.0-11.4	CABS	

7

Temperature

Lesley Lloyd Haire

7

Temperature

Lesley Lloyd Haire

National Institute for Medical Research,
London, U.K.

7.1. Temperature as a Crystallization Parameter

Many proteins vary in solubility as a function of temperature. Protein solubility usually increases with increasing temperature in low ionic strength conditions such as PEG or MPD solutions. This is generally referred to as **normal** solubility. At high ionic strength, proteins are generally less soluble at 25°C than at 4°C, and this is known as **retrograde** solubility.

Crystallization has been reported to occur for proteins over a whole range of temperature, from less than 0°C to around 60°C, although it is usually conducted at either 4°C or at room temperature. Low temperature is advantageous if the protein is heat labile, as it helps stabilize the protein and inhibit microbial growth.

At low ionic strength (i.e., with no precipitant present) some proteins are extremely sensitive to temperature, providing a means by which they can be crystallized. The production of crystals during an initial concentration step by centrifugation through an ultrafiltration device (Centricon microconcentrator) was described by Pitts.[1]

I have observed crystals of several proteins, including cholesterol oxidase, *Thermoanaerobacterium thermosulfurigenes* xylose isomerase, and pig collagenase during concentration at 4°C (L. Lloyd Haire, unpublished results). These crystals were not all suitable for

X-ray diffraction studies, but subsequent screening with precipitants yielded diffracting crystals.

7.2. Recommendations When Working at 4°C

It is advisable when screening with organic solvents (e.g., volatile organics such as isopropanol, ethanol, or nonvolatile organics such as PEG 400 or MPD) to set up the trials at 4°C. Generally, protein solubility decreases with a decrease in temperature in the presence of such solvents. The protein may denature in the presence of organic solvents and this may be minimized by working at low temperature.

Condensation on cover slips may be a problem at low temperature. This can be overcome in several ways:

- A dummy plate with water in the wells may be placed above and below a stack of trays.
- Some salt, e.g., 4 M NaCl, may be added to the reservoir (in addition to its contents) to remove the excessive moisture from the cover slip. This should be added in increments of 50 µl over a period of days.
- The sitting drop method may be used instead of hanging drops.

It is possible to perform 4°C crystallization experiments at room temperature and avoid long periods in the cold room. Samples and solutions are pre-equilibrated at 4°C and the crystallization tray placed on ice. When the drops have been set up, the tray is transferred to a 4°C incubator or cold room. If the trays are stored in a cold room, it is a good idea to place them inside a suitably sized polystyrene box to minimize drafts and temperature fluctuations which may be caused by excessive air movement due to other workers in the area.

7.3. Testing the Effects of Temperature

Temperature is often overlooked when screening crystallization variables. In order to test the effect of temperature on a protein, try setting up duplicate crystallization screens and conduct parallel experiments at two temperatures, e.g., 4°C and 18°C. If differences in the solubility behavior of the protein are observed, e.g., a clear drop in one condition at 4°C and precipitate in the duplicate drop at 18°C, this indicates that temperature is affecting the solubility and should be explored further as a crystallization variable. If protein is only available in limited supply and it is not feasible to set up duplicate screens at different temperatures, crystallization trials can be set up at one temperature, e.g., 4°C and left 1-4 weeks. The results are then noted and the tray transferred to a different temperature, e.g., 18°C, and again left for 1-4 weeks. This procedure may be repeated again at a higher temperature. Comparison of the results at the different temperatures will reveal whether the solubility of the protein is temperature dependent. If so, then use temperature as a variable during optimization.

7.4. Crystallization by Temperature Gradients

In some cases temperature gradients have been used to optimize protein crystallization. A classic example is that of insulin where a "hot box" technique (as used by Dr. G. Dodson) was used to produce a temperature gradient of solubility.[2] The protein was dissolved in a buffer at 50°C (without precipitant), placed in an insulated container, and allowed to cool slowly. This resulted in a decrease in the solubility of the protein with a corresponding increase in supersaturation. In this way the approach to supersaturation at lower temperature was controlled. Other examples of proteins that show a sharp decrease in solubility as a function of temperature include porcine pancreatic elastase and α-amylase. Elastase was crystallized from 50 mM sodium citrate, pH 5.5, 9 mg/ml protein, by lowering the temperature from 25 to 20°C.[3] Crystals of α–amylase were grown from a solution containing 5 mM $CaCl_2$ at pH 8 in low ionic strength buffer by lowering the temperature from 25 to 12°C.[4] The same principle can be applied to other proteins, adjusting the initial and final temperatures according to the variation of protein solubility with temperature, with or without precipitant as appropriate.

References and Further Reading

1. Pitts JE: **Crystallization by centrifugation.** *Nature* 1992, **355**:117.

   X-ray quality crystals of an aspartic proteinase appeared in the Amicon ultrafiltration device during concentration by low speed centrifugation.

2. Blundell TL, Johnson LN: *Protein Crystallography.* New York: Academic Press 1976:59-82.

   Description of the "hot box" technique used to crystallize insulin.

3. Shotton DM, Hartley BS, Camerman H, Hofmann T, Nyborg SC, Rao L: **Crystalline porcine pancreatic elastase.** *J Mol Biol* 1968, **32**:155-156.

   Use of temperature shift as a crystallization method.

4. McPherson A, Rich A: **X-ray crystallographic analysis of swine pancreas -amylase.** *Biochem. Biophys.Acta* 1972, **285**:493-497.

   Another example of crystallization by temperature shift.

8

Crystallization Strategies

Terese Bergfors

8

Crystallization Strategies

Terese Bergfors

Uppsala University, Uppsala, Sweden

> *"Plans are nothing, planning is everything."*
> ——*Dwight D. Eisenhower,*
> *U.S. general and president*

The crystallization of proteins can be divided into two stages: 1) initial screening to obtain any kind of crystals or promising precipitates, and 2) optimization to improve the crystals. The strategies for these two objectives will be different.

That is as far as everyone agrees. What people don't agree on is which strategies should be used, especially for the initial screening.

8.1. The Problem

The problem in designing a crystallization screen lies in the large number of parameters that affects crystallization. What is the most effective strategy for sampling them? Should one or two parameters be varied at a time, keeping the others constant, or should all the parameters be varied simultaneously in a random way? How can one separate the single effects of any two parameters from their interactive effect in the crystallization drop?

The difficulty in designing an optimal crystallization screen is often further complicated by limitations on the stability and availability of the protein.

8.2. Types of Screens: Pros and Cons

Several strategies have been proposed for dealing with a sampling problem of this enormity. Which one is best, i.e., which of these strategies gives the best crystals the fastest, with the least amount of protein, and work?

It is difficult to rigorously compare the efficiency of the different crystallization strategies, but at least one attempt to do so has been made.[1] In this study of five test proteins, a **random screen** was compared with two other screen types, **grid** and **footprint**. The random screen turned out to be the most efficient of the three, i.e., it produced crystals in the least number of trials.

Sparse matrix screens are frequently employed for initial screening, followed by **grid screens** to refine the promising conditions. The sparse matrix design was first introduced in 1991 by Jancarik and Kim[2] and is also sometimes referred to as a **"fast screen."** It is now commercially sold by Hampton Research under the name **Crystal Screen.** (See Appendix A5.)

Sparse matrix screens are biased samplings of crystallization parameters, selected from known or published successful crystallization conditions (usually from a database). Thus, they are heavily biased towards conditions which have worked previously and are therefore not truly random (in the sense that statisticians use the word). So long as they are successful in producing some initial crystals, this is not a problem. However, if no "hits" are produced, the amount of information that can be extracted from the sparse matrix screen is limited, since it is not statistically balanced.

Statisticians who work with large sampling problems recommend the **incomplete factorial** design as the most efficient. In this design, the parameters are all varied simultaneously and randomly.[3] Previous knowledge may affect the choice of parameters and their levels, as with sparse matrix designs. However, in the incomplete factorial design all the parameters and their levels (concentrations) are balanced, so that the conclusions from the results are statistically significant.

As an example, let us say that you are working with a metal-binding protein that is stable in the pH range 4-6. This previous knowledge will lead you to select a particular pH range and the category "divalent cations" for testing. In the Jancarik and Kim sparse matrix screen, there are 38 experiments without any divalent cations and 9 with them (Ca^{2+}, Mg^{2+}, and Zn^{2+}). Thus, the parameter "divalent cations" is not balanced. If this screen is used but no crystals are found, it is impossible to say if the effect of cations is statistically significant or not. By contrast, an incomplete factorial design would balance the number of experiments with and without different cations.

The incomplete factorial design is not simple, and the convenience of the sparse matrix screen, especially in the form of ready-made kits, contributes to the latter's greater popularity.

Grid screens (see Chapter 5) select one or two parameters to vary while keeping the others constant. The results are easy to interpret and give information even if no crystals are forthcoming. The disadvantage of grid screens (as an initial screening strategy) is that generally they require more trials to find "hits." However, grid screens were used quite routinely until the advent of the first sparse matrix design in 1991.

The four strategies presented in the following chapters, have been chosen on the basis of diversity:

Strategy 1: **a flexible sparse matrix** design by Johan Zeelen. This is a sparse matrix design like the Crystal Screens but with 48 unique conditions. It is flexible in the sense that the solutions are not pre-formulated. Other precipitants or buffers can be substituted if more appropriate for your particular protein.

Strategy 2: Madeleine Ries-Kautt's method is based on the effect of the **Hofmeister series** depending on the net charge of the protein.

Strategy 3: Enrico Stura's **reverse screen** is based on the determination of the specificity of precipitants and buffers. The initial search is done at high levels of supersaturation followed by a second pass with additives.

Strategy 4: The Imperial College screen is an example of a **systematic grid screen**, covering 60 conditions.

In addition, the following screens are worth the reader's attention:

1. CRYSTOOL (a program for generating a custom **random** screen).[1]

2. IFDesign (a program for the generation and analysis of incomplete factorial experiments).[4]

3. **Response surface methods** (a **factorial** design for optimization).[5]

4. Footprint screening (a solubility footprint of the protein, similar to a grid screen).[6]

5. Wizard 96 (a **sparse matrix** screen with 96 unique conditions, i.e., conditions not included in any other sparse matrix screens).[7] See Appendix A5 for the conditions in this and other commercially available sparse matrix screens.

6. **Grid screens** (see the Chapter 5, *Precipitants* and references therein).

7. Two **other approaches** not included in these other categories.[8]

8.3. Crystallization Strategy is More Than a Choice of Screening Method

An overall strategy should include the handling and characterization of the macromolecule. Ultimately, the outcome of the crystallization experiment depends more on the state of material than screen type. A "crystallizable" protein will eventually crystallize, if the researcher persists. The macromolecule has many different possibilities to form the necessary crystal contacts depending on its net charge, nature of the precipitant, state of hydration, etc. There is not one **unique** crystallization condition; the same macromolecule can crystallize in different crystal forms.

On the other hand, a "badly behaved" protein, i.e., one incapable of making the necessary protein-protein contacts for nucleation, will never crystallize regardless of how many statistically sound screens are employed. Any number of factors can be responsible for the inability of the protein to form proper contacts, such as

- glycosylation
- proteolytic degradation

- aggregation
- conformational heterogeneities
- interference from His-tags, floppy N- or C-termini
- oxidation of cysteines, etc.

Reference[9] is a review article with some good examples of protein engineering to get crystals. Modifications to or further purification of the protein should always be considered if it seems intractable to crystallization.

The flow charts in references[10-13] can be useful guides in planning an overall strategy for crystallization.

References and Further Reading

1. Segelke B: **Computing of a high efficiency screening protocol for crystals.** *Proceedings of the ICCB M VII,* 1998, Granada, Spain.

   See http://www-structure.llnl.gov/crystool for the CRYSTOOL program.

2. Jancarik J, Kim S-H: **Sparse matrix sampling: a screening method for crystallization of proteins.** *J Appl Cryst* 1991, **24**:409-411.

   The original sparse matrix screen.

3. Carter C: **Experimental design, quantitative analysis, and the cartography of crystal growth.** In *Crystallization of Nucleic Acids and Proteins: A Practical Approach.* Edited by Ducruix A and Giegé R. IRL/Oxford Press; in press.

   Incomplete factorials in theory and practice.

4. Knight S: **IFDesign** Version 1.5 Users manual.

   Program to generate incomplete factorial screens.

   Contact: stefan@xray.bmc.uu.se.

5. Carter C: **Response surface methods for optimizing and improving reproducibility of crystal growth.** *Meth Enzymol* 1997, **276**:74-99.

   Optimization by use of factorials. Not an easy read.

6. Stura EA, Nemerow GR, Wilson IA: **Strategies in the crystallization of glycoproteins and protein complexes.** *J Cryst Growth* 1992, **122**:273-285.

   See: http://www.scripps.edu/~stura/cryst/screen.html. Footprint screening.

7. The Wizard Matrices were developed by Steve L. Sarfaty and Wim G.J. Hol at the University of Washington (Seattle, U.S.A., 1998). Emerald BioStructures, Inc. has obtained an exclusive license from the University of Washington to market the Wizard I and II (U.S. Patents Pending). Contact: emerald_biostructures@rocketmail.com.

8. McPherson A: **Two approaches to the rapid screening of crystallization conditions.** *J Cryst Growth* 1992, **122**:161-167.
    Additional screening approaches.

9. Forest K, Schutt C: **Protein engineering for structure determination.** *Curr Opin Struct Biol* 1992, **2**:576-581.
    Review article with examples of modifications made to proteins in order to get them to crystallize.

The following contain flow charts:

10. **Crystal Growth 101: Optimization flow chart for macromolecular crystallization.** Technical bulletin # CG1011 Hampton Research, 27632 El Lazo Road, Suite 100, Laguna Niguel, California 92677-3913, USA.
    Flow chart.

11. McPherson A: **Current approaches to macromolecular crystallization.** *Eur J Biochem* 1990, **189**:1-23.
    Two extensive flow charts, recommended for the beginner.

12. D'Arcy A: **Crystallizing proteins: a rational approach?** *Acta Cryst* 1994, **D50**:469-471.
    Flow chart.

13. Gilliland GL, Bickham DM: **The biological macromolecule crystallization database: a tool for developing crystallization strategies.** *Methods: Companion to Meth Enzymol* 1990, **1**:6-11.
    Flow chart based on the BMCD.

9

Strategy 1: A Flexible Sparse Matrix Screen

Johan Philip Zeelen

9

Strategy 1: A Flexible Sparse Matrix Screen

Johan Philip Zeelen

Max Planck Institute of Biophysics,
Frankfurt am Main, Germany

9.1. The Protein Sample

Crystallographic structure determinations start with "crystal-quality" protein in sufficient amounts (>10 mg/batch). A good indication of protein purity can be obtained by SDS PAGE (sodium dodecyl sulfate polyacrylamide gel electrophoresis) and staining with Coomassie. To check whether the protein is homogeneous, run an IEF (iso-electric focusing) gel and size exclusion chromatography. If possible apply mass spectrometry and dynamic light scattering to check for impurities and monodispersity. If large amounts are available but the protein is not homogeneous, change the purification protocol or add an extra purification step. When the sample contains minor impurities or only a few milligrams of protein are available, try to find initial conditions for crystallization. Further purification can be performed later to optimize the crystals. It is better to try to crystallize a reasonably pure and homogeneous protein batch than to have an extremely pure protein but not enough to crystallize.

9.1.1. Protein Concentration and Starting Buffer

The recommended protein concentration is 5-20 mg/ml, measured with a Bradford assay using BSA (bovine serum albumin) as reference.

Dialyze the protein in a small dialysis bag against 1 liter starting buffer. The salt concentration in the starting buffer should be as low as possible. If the protein is not soluble, change the pH of the buffer or add small amounts of different salts. The buffer concentration in the starting solution (5-10 mM) is 10-20 times lower than the buffer concentration used in the well (100 mM in the first screen). To maintain homogeneity during storage and crystallization, stabilize the protein with DTT (dithiothreitol), EDTA (ethylenediaminetetraaceticacid), and protease inhibitors. Add NaN_3 to prevent bacterial growth. A typical protein buffer is 10 mM TEA/HCl (triethanolamine), pH 7.5, 25 mM NaCl, 1 mM DTT, 1 mM EDTA, and 1 mM NaN_3.

Depending on the stability of the protein, store in the refrigerator or in small aliquots in the freezer. Check protein stability after storage, and centrifuge or filter the protein solution before use to remove precipitated protein.

9.2. Crystallization with a "Fast-Screening" Protocol

Protein crystallization is carried out in several steps:

1. Crystallization/nucleation conditions or promising leads are identified with the initial screen.
2. The conditions from the initial screen are fine-tuned in an adjusted screen. Usually this means refining the parameters so that supersaturation is achieved more slowly.
3. Large single crystals (> 0.2 mm) suitable for X-ray diffraction (better than 3Å resolution) are obtained in an optimization screen based on the results found in the initial and adjusted screens.

6.2.1. The Initial Screen

The "sparse matrix screen" is a way to screen a large number of parameters with a limited amount of protein. A coarse matrix can be used because it has been observed that nucleation can occur over a broad range of parameters. In the initial screen we examine the roles of pH, precipitant, and temperature. To simplify the interpretation of the results, the crystallization conditions are sorted by precipitant. An example of an initial screen is given in Table 9.1. Although the preparation of wells from the stock solutions is more labor intensive than using ready-made solutions, it has the advantage that precipitant concentration, pH, and additive can be manipulated independently, hence the name "flexible" sparse matrix screen. Because the same conditions are used in each initial screening, this approach is well suited for automation.

9.2.2. Preparing the Initial Screen

1. Prepare the stock solutions, consisting of different buffers, precipitants, and additives, in 50 ml Falcon tubes, as indicated in Table 9.2. Filter trough a 0.22 μm filter and store at room temperature.

2. Wash cover slips (22 mm round or 20 mm square) with ethanol; dry and siliconize with a siliconizing agent to make them hydrophobic.

3. Label two 24-well tissue culture plates (e.g., Linbro plates, ICN cat. no. 76-033-05) and place a piece of tape or plasticine in each corner. This is to prevent the lid from crashing into the cover slips if several trays are stacked on top of each other.

4. Apply grease or high-viscosity oil around the rim of the wells.

5. Pipette the stock solutions into the different wells as indicated in Table 9.2. The end volume should be 1 ml. Carefully mix the wells by shaking the plate.

6. Pipette 1 μl protein solution and mix with 1 μl well solution on a cover slip, using a Gilson Pipetman P2.

7. Invert the cover slip and seal it over the well chamber. (Use Millipore tweezers cat. no.XX62 000 06.)

8. Prepare two sets of these 48 conditions: Place one set at 4°C and one at room temperature. Cool the 4°C plates before making the drops.

9. Examine the screens immediately and then every day (in the first week) to follow the precipitation/crystallization of the protein over time.

9.2.3. Adjusted Screen

Repeat all wells where the initial screen does not indicate nucleation, with the following modifications. For those conditions where the protein precipitated immediately after mixing, halve the precipitant concentration. For conditions without visible precipitate after more than two weeks, increase/double the protein or precipitant concentration. Repeat until the precipitation point for all wells is determined. (I once found crystals when the precipitant concentration was reduced to 25 mM.) There is no need to repeat the wells with protein precipitates that formed between two days and two weeks.

9.3. Optimize the Crystallization Conditions

9.3.1. Optimization Step 1

The conditions that favored nucleation in the initial screens are now used in the optimization screen. For each condition, a new tray is set up in which one parameter is varied and the others kept constant:

Row A   different *precipitant 1* concentrations **with** *precipitant 2*

Row B   different *precipitant 1* concentrations **without** *precipitant 2*

Row C   buffer and pH (pH range 4.5-9.5) varied

Row D   protein concentration varied

The range for the precipitant in the well is usually between half and full strength of the concentration which gave nuclei/crystals. The rate at which the nucleation occurred is used to determine this. The concentration range of *precipitant 1* is the same in rows A and B. (If the condition from the initial screen being optimized only uses one precipitant, skip row A.)

In the example in Table 9.3, two precipitants were used: 25% PEG 6000 and 200 mM Li_2SO_4. If the nucleation occurred after 3 days, the range of PEG 6000 to be chosen in optimization should be 12.5-25%. If it took 1-2 weeks, use a higher range (20-30%). Use a lower range for nucleation within 2 days (10-20%). Comparing the results from row A with B will show whether *precipitant 2* is essential. The example given shows that Li_2SO_4 is essential because crystals are found only in row A, at a PEG 6000 concentration between 20 and 25%.

In row C we change the buffer but keep the precipitant concentrations constant. The result from row C will show the pH dependence. In our example, at low pH the protein precipitates (possibly denatured) and the range between 6.5 and 8.5 gives plates and crystals. In row D, the well condition that gave crystals in the initial screen is kept constant but the protein concentration is varied. Before the drops are prepared, dilute the protein with the same buffer used for the protein dialysis. The example shows that a protein concentration of 4-5 mg/ml is better then 6 mg/ml. Comparing the identical conditions A6, C3, and D6 in this example with the condition 2D5 in the initial screen will give an indication of the reproducibility of the experiment. If the results from the comparable drops are not identical, check the protein stability.

9.3.2. Optimization Step 2

The next step is a broad grid screen with variation in pH (in steps of 0.5 pH units) against precipitant, refining the parameters until the optimal crystallization conditions are found. Remember: It is not size or morphology but the diffraction quality of the crystals that is important. Looking at the results from Table 9.3, the first broad grid screen would test pH 6.0, 6.5, 7.0, 7.5, 8.0, 8.5 versus 19%, 21%, 23%, and 25% PEG 6000. In addition all wells contain 200 mM Li_2SO_4 and 1 mM DTT, 1 mM EDTA, and 1 mM NaN_3. Protein concentration is 5 mg/ml. A second broad screen would cover the same pH range versus 50 mM, 100 mM, 150 mM, and 200 mM Li_2SO_4 plus 25% PEG 6000 and 1 mM DTT, 1 mM EDTA, and 1 mM NaN_3 in the wells.

9.3.3. If the Crystals Are Not Good Enough

Add inhibitors, substrate analogues, or additives (Table 9.4) to the protein drop, using the optimized well conditions. Repeat the initial screen with the compound if an effect is found. Also check the initial and adjusted screens for other conditions that were not optimized before.

9.3.4. When Nucleation Is Not Evident

After screening and adjusting the composition of all wells, look for inhibitors and substrate analogues to stabilize the protein. Check and improve the purity and homogeneity of the protein and try a different initial screen.

References and Further Reading

1. Zeelen JP, Hiltunen JK, Ceska TA, Wierenga RK: **Crystallization experiments with 2-enoyl-coA hydratase, using an automated 'fast-screening' crystallization protocol.** *Acta Cryst* 1994, **D50**:443-447.

   This article describes the development of the flexible sparse matrix screen and use of a crystallization robot.

2. Radha Kishan KV, Zeelen JP, Noble MEM, Borchert TV, Wierenga, RK: **Comparison of the structures and the crystal contacts of trypanosomal triosephosphate isomerase in four different crystal forms.** *Protein Science* 1994, **3**:779-787

   The protein crystallized in four different space groups with different resolution.

A 9.1. Tables for the "Flexible Sparse Matrix Screen"

Table 9.1. Composition of the 48 well solutions of the initial screen

	buffer * 100 mM	pH **	precipitant 1	precipitant 2
1A1		8.2	1.75 M $(NH_4)_2HPO_4$	
1A2	ADA	6.5	2.0 M $(NH_4)_2SO_4$	100 mM $MgSO_4$
1A3	Citrate	5.5	2.0 M $(NH_4)_2SO_4$	200 mM NaCl
1A4	CHES	9.5	2.0 M $(NH_4)_2SO_4$	5 % MPD
1A5	TEA	7.5	2.0 M $(NH_4)_2SO_4$	2 % PEG 400
1A6	ADA	6.5	2.0 M $(NH_4)_2SO_4$	10 % Ethanol

Table 9.1—*Continued*

	buffer * 100 mM	pH **	precipitant 1	precipitant 2
1B1	Acetate	4.5	2.5 M $(NH_4)_2SO_4$	200 mM Li_2SO_4
1B2	TRIS	8.5	3.0 M $(NH_4)_2SO_4$	
1B3	TEA	7.5	30 % t-Butanol	
1B4	ADA	6.5	20 % Ethanol	200 mM KCl
1B5	Citrate	5.5	30 % Ethanol	
1B6	Acetate	4.5	30 % Ethanol	10 % PEG 6000
1C1	TRIS	8.5	40 % Ethanol	200 mM $MgCl_2$
1C2	TEA	7.5	10 % Isopropanol	20 % PEG 400
1C3	Acetate	4.5	20 % Isopropanol	200 mM KCl
1C4	Citrate	5.5	20 % Isopropanol	2 % PEG 1500
1C5	CHES	9.5	30 % Isopropanol	200 mM $MgSO_4$
1C6		5.5	1.5 M K_2HPO_4/NaOH	
1D1	TEA	7.5	1.5 M KNa Tartrate	
1D2	Acetate	4.5	1.0 M Li_2SO_4	
1D3	TRIS	8.5	1.0 M Li_2SO_4	1.0 M $(NH_4)_2SO_4$
1D4	TEA	7.5	1.5 M Li_2SO_4	
1D5	Citrate	5.5	1.5 M $MgSO_4$	
1D6	TRIS	8.5	2.0 M $MgSO_4$	
2A1	CHES	9.5	20 % MPD	200 mM $MgCl_2$
2A2	Citrate	5.5	30 % MPD	
2A3	TRIS	8.5	30 % MPD	2.5 % t-Butanol
2A4	Acetate	4.5	40 % MPD	
2A5	TEA	7.5	40 % MPD	200 mM NaCl
2A6	CHES	9.5	1.0 M Na Citrate	
2B1	TEA	7.5	1.5 M Na Citrate	
2B2	TRIS	8.5	2.0 M NaCl	
2B3	TEA	7.5	2.0 M NaCl	10 % PEG 400
2B4	Acetate	4.5	3.0 M NaCl	
2B5	Acetate	4.5	2.0 M Na Formate	
2B6	Citrate	5.5	20 % PEG 400	200 mM KCl
2C1	TEA	7.5	30 % PEG 400	200 mM $MgCl_2$
2C2	TRIS	8.5	40 % PEG 400	
2C3	TRIS	8.5	15 % PEG 1500	5 % MPD
2C4	Citrate	5.5	20 % PEG 1500	
2C5	ADA	6.5	20 % PEG 1500	200 mM KCl
2C6	CHES	9.5	25 % PEG 1500	200 mM $MgSO_4$

Table 9.1—*continued*

	buffer * 100 mM	pH **	precipitant 1	precipitant 2
2D1	TEA	7.5	30 % PEG 1500	200 mM $(NH_4)_2SO_4$
2D2	Acetate	4.5	20 % PEG 6000	1.0 M NaCl
2D3	Citrate	5.5	20 % PEG 6000	2.5% t-Butanol
2D4	TEA	7.5	25 % PEG 6000	
2D5	ADA	6.5	25 % PEG 6000	200 mM Li_2SO_4
2D6	Acetate	4.5	30 % PEG 6000	

* Each well contains 1 mM DTT, 1 mM EDTA and 1 mM NaN_3 because they have been added previously to the stock solutions.

** The pH is the pH of the original buffer, as specified in Table 9.2. It should be noted that the actual pH in the well solution can be significantly different due to the various additions, or the change in temperature.

MPD = 2-Methyl-2,4-pentanediol, PEG = Polyethylene glycol,
ADA = N-[2-Acetamido]-2-iminodiacetic acid, TEA = Triethanolamine,
TRIS = tris[Hydroxymethyl]aminomethane, and
CHES = 2-[NN-Cyclohexylamino]ethanesulfonic acid.

Table 9.2. Stock solutions and pipetting scheme for the initial screen

Make 50 ml of the 25 stock solutions and add DEN* to each. Alternatively, add 10 µl 100 mM DEN* to each well, and reduce all volumes of added water with 10 µl. The pipetting scheme below is based on two trays of 24 wells each. Pipette the indicated volumes in the corresponding wells. The total volume in the well is 1 ml. Work carefully to prevent mistakes.

Stock solution	Tray 1 (Volumes in µl)						Tray 2 (Volumes in µl)					
1. H_2O	125	289	289	229	289	229	400	300	275	100	60	344
	75	43	600	500	600	350	67	500	300	300	400	300
	400	400	500	650	520	250	200	100	425	400	200	195
	150	344	58	67	300	100	93	200	375	275	164	150
2. 50 % MPD				100			400	600	600	800	800	
									100			
3. 50 % PEG 400					40							
									200			400
		400					600	800				

Table 9.2—*continued*

Stock solution	Tray 1 (Volumes in µl)						Tray 2 (Volumes in µl)				
4. 40 %											
PEG 1500											
				50				375	500	500	625
							750				
5. 40 %											
PEG 6000						250					
							500	500	625	625	750
6. 1.0 M Acetic								100			
acid/NaOH	100				100			100	100		
			100								
pH = 4.5		100					100				100
7. 1.0 M Citric			100				100				
acid/NaOH					100						100
				100					100		
pH = 5.5					100			100			
8. 1.0 M		100			100						
ADA/NaOH			100								
										100	
pH = 6.5										100	
9. 1.0 M				100							100
TEA/HCl			100				100	100			
		100					100				
pH = 7.5	100			100			100		100		
10. 1.0 M								100			
TRIS/HCl		100					100				
	100						100	100			
pH = 8.5			100		100						
11. 1.0 M				100			100				100
CHES/NaOH											
					100						100
pH =9.5											
12. 1.0 M KCl											
				200							200
			200							200	
13. 2.0 M	875										
$(NH_4)_2HPO_4$											
pH= 8.2											
14. 3.5 M		571	571	571	571	571					
$(NH_4)_2SO_4$	714	857									
			286				57				

88   *Zeelen*

Table 9.2—*continued*

Stock solution	Tray 1 (Volumes in µl)					Tray 2 (Volumes in µl)				
15. 1.8 M Li$_2$SO$_4$	111									
		556	556	833					111	
16. 2.0 M MgCl$_2$						100				
	100					100				
17. 2.5 M MgSO$_4$		40								
				80						80
				600	800					
18. 2.0 M KH$_2$PO$_4$/NaOH pH = 5.5										
				750						
19. 2.0 M KNaTartrate										
	750									
20. 1.8 M Na$_3$Citrate										556
						833				
21. 5.0 M NaCl		40							40	
						400	400	600		
						200				
22. 4.0 M Na Formate									500	
23. 100 % tert-Butanol								25		
		300								
								25		
24. 100 % Ethanol					100					
			200	300	300					
	400									
25. 100 % Isopropanol										
		100	200	200	300					

DEN = 1 mM Dithiothreitol, 1 mM Ethylenediaminetetraacetic acid (EDTA), and 1 mM NaN$_3$

MPD = 2-Methyl-2,4-pentanediol, PEG = Polyethylene glycol, ADA = N-[2-Acetamido]-2-iminodiacetic acid,

TEA = Triethanolamine, TRIS = tris[Hydroxymethyl]aminomethane, and

CHES = 2-[NN-Cyclohexylamino]ethanesulfonic acid.

Table 9.3. Example of an optimization experiment

Condition from initial screen, with small crystals after 3 days:
100 mM ADA/NaOH, pH = 6.5, 25% PEG 6000, 200 mM Li_2SO_4
1 mM DTT, 1 mM EDTA, and 1 mM NaN_3

			Result *:	
Date: 29-01-'97 Temperature: 20°C. Experiment number: 96025-1				
Experiment name: *E. coli* Triose phosphate isomerase.				
Protein	: E-TIM batch 96076			
[protein]	: 6 mg/ml			
volume	: 2 µl			
Mother liquor	: 2 µl			
Buffer	: 10 mM TEA/HCl pH=7.6, 20 mM NaCl, 1 mM phosphoglycolate, DEN			
Mother liquor: A		A1. 12.5%	0 0 0 0	clear
Buffer	: 100 mM ADA/NaOH	A2. 15.0%	0 0 0 0	clear
pH	: 6.5	A3. 17.5%	0 0 0 0	clear
precipitants	: PEG 6000	A4. 20.0%	0 0 9 9	small crystals
Variations in	: [PEG 6000]	A5. 22.5%	0 0 9 9	small crystals
Additions	: 200 mM Li_2SO_4, DEN	A6. 25.0%	0 6 6 6	micro crystals
Mother liquor: B		B1. 12.5%	0 0 0 0	clear
Buffer	: 100 mM ADA/NaOH	B2. 15.0%	0 0 0 0	clear
pH	: 6.5	B3. 17.5%	0 0 0 0	clear
precipitants	: PEG 6000	B4. 20.0%	0 0 4 4	particulated precipitate
Variations in	: [PEG 6000]	B5. 22.5%	0 0 4 4	particulated precipitate
Additions	: DEN	B6. 25.0%	0 0 4 4	particulated precipitate
Mother liquor: C		C1. pH=4.5	3 3 3 3	heavy precipitate
Buffer	: 100 mM	C2. pH=5.5	0 4 4 4	particulated precipitate
precipitants	: 25% PEG 6000	C3. pH=6.5	0 6 6 6	micro crystals
Variations in	: [pH]	C4. pH=7.5	0 8 8 8	plates
Additions	: 200 mM Li_2SO_4, DEN	C5. pH=8.5	0 0 6 6	micro crystals
		C6. pH=9.5	0 0 0 4	particulated precipitate
Mother liquor: D		D1. 1 mg/ml	0 0 0 0	clear
Buffer	: 100 mM ADA/NaOH	D2. 2 mg/ml	0 0 0 0	clear
pH	: 6.5	D3. 3 mg/ml	0 0 0 0	clear
precipitants	: 25% PEG 6000	D4. 4 mg/ml	0 0 0 9	crystals
Variations in	: [protein]	D5. 5 mg/ml	0 0 0 9	crystals
Additions	: 200 mM Li_2SO_4, DEN	D6. 6 mg/ml	0 9 9 9	small crystals

Phosphoglycolate = a substrate analogue which fixes a flexible loop in a 'closed' con-
formation. TEA/HCl = triethanolamine pH adjusted with HCl, ADA/NaOH =
N-[2-acetamido]-2-iminodiacetic acid pH adjusted with NaOH, DEN = 1 mM DTT,
1 mM EDTA, and 1 mM NaN_3

* See Chapter 13, *Interpretation of Results* for an explanation of the numbers.

Table 9.4. Additives

27 additives for optimization. The crystals can also be further improved by adding substrate analogues, inhibitors, ligands, or other additives not included here.

5 % Jeffamine	100 mM $(NH_4)_2SO_4$	100 mM 1,6-hexanediol
5 % polypropyleneglycol P400	100 mM CsCl	100 mM dextran sulfate
5 % PEG 400	100 mM $CoSO_4$	100 mM 6-aminocaproic acid
5 % ethyleneglycol	100 mM $MnCl_2$	100 mM 1,6-hexanediamine
5 % 2-methyl-2,4-pentanediol	100 mM KCl	100 mM 1,8-diaminooctane
5 % glycerol	100 mM $ZnSO_4$	100 mM spermidine
5 % dioxane	100 mM $LiCl_2$	100 mM spermine
5 % dimethyl sulfoxide	100 mM $MgCl_2$	0.17 mM n-dodecyl-β-D-maltoside
5 % n-octanol	100 mM glucose	25 mM n-octyl-β-D-glucopyranoside

10

Strategy 2: An Alternative to Sparse Matrix Screens

Madeleine Riès-Kautt

10

Strategy 2: An Alternative to Sparse Matrix Screens

Madeleine Riès-Kautt

LEBS, CNRS Gif-sur-Yvette, France

10.1. Introduction

When you use a sparse matrix screen, you are trying to check if a crystallization condition which has already produced crystals with a different protein could work with your protein. By analogy, would you try to solve a structure by testing all the proteins in the PDB for molecular replacement? You may get lucky, but if it fails you have no other choice but to change strategy.

The strategy proposed here is based on physico-chemical considerations.[1] This means that we consider:

- the *net charge of the protein* (which is approximately zero around the pI, positive at pH < pI and negative at pH > pI).
- the *counter-ions* (i.e., ions of opposite charge). As soon as the protein is charged, counter-ions will be present to maintain the electroneutrality of the solution.

The effects of increasing the salt concentration of the protein solution may be summarized as follows. There is:

- a change in the hydration of the protein, because the ions themselves need water for solvation.
- a decrease in the repulsive electrostatic protein-protein interactions, allowing the macromolecules to come closer to each other.

- direct interaction of ions with residues at the surface of the protein to form a *protein salt*. Just as the solubility and the crystallization conditions are different for calcium phosphate and sodium phosphate, likewise are they different for lysozyme chloride and lysozyme nitrate.

Our approach, described more extensively in reference[2], is to find the right conditions for macromolecules to come into close contact (before aggregating as an amorphous solid), without precluding the opportunity for these macromolecules to build a three-dimensional lattice, i.e., a crystal. The latter still remains unpredictable and requires navigation in a phase diagram to search the nucleation range.

The search for crystallization conditions is done stepwise:

1. estimation of the net charge over a large pH range, and choice of the pH value(s) of crystallization.
2. search of the limit of nucleation/precipitation by rapid tests to locate the range of crystallizing agent to focus on.
3. refinement of conditions by small increases in the concentration of crystallizing agent in order to promote crystals, or at least any solid phase. The crystallizing agents are chosen depending on the protein net charge.
4. adjustment of the conditions where some solid phase appears (spherulites, microcrystals, etc.) to obtain large single crystals.

10.2. Estimation of the Net Charge

The net charge of a macromolecule, Z_p, is the difference between the number of positive and negative charges arising from the deprotonation of acidic groups and the protonation of basic groups. The pKa values of solvent-accessible charged residues taken for this calculation are: 3.5 (COOH-terminal), 4.5 (Asp, Glu), 7.6 (NH_2-terminal), 6.2 (His), 9.5 (Tyr), 9.3 (free Cys), 10.4 (Lys), and 12.0 (Arg).

The Z_p at different pH values can be calculated in different ways:

a. On the World Wide Web at:

http://www-biol.univ-mrs.fr/d_abim/compo-p.html
http://www.expasy.ch/sprot/protparam.html
http://www.infobiogen.fr/service/deambulum

These give the calculated pI, and the shape of the net charge variation with pH. For a more precise calculation use **b** or **c**.

b. Complete the spread sheet given in appendix A 10.1 of this chapter.

The charge contribution z of one charged residue is given in each column for the indicated pH value and type of residue.

Write in each box of the first row the number of each type of amino acid contained in your protein. For each column, multiply the indicated contribution of one residue by this number. Sum all charges for a given row to obtain the net charge at this pH.

c. Use programs.

Use programs such as Kaleidagraph, Excel, etc., to calculate automatically a table equivalent to what is given in appendix A 10.1. Equations are given in appendix A 10.2.

N.B. The calculated pI (pH for a net charge of zero) should be confirmed by the electrophoretic measurement of the experimental pI. If a difference is observed between the estimated pI and the experimental one, it means that the pKa of some residue(s) is different from the average values used for the calculation (due to complexation, salt bridges, etc.).

10.3. Choice of the pH for Crystallization

The higher the net charge of the protein, the higher its solubility will be. As described in section 10.7, the best crystallization conditions are found at high solubility, rather than low (near the pI).

Ideally it is recommended to test crystallization at three pH values, where the protein bears a positive, negative and zero net charge, respectively. This depends of course on what is known about the stability of the protein and on the experimental feasibility.

If the pI is lower than 4 (or higher than 9), only the state of negative (or respectively positive) net charge can be tested. Choose two pH values for which the net charge is significantly different. Note that the most important changes of the net charge occur around the pKa values of the acidic (pH ≈ 4.5) and basic (pH > 9) residues. Between pH 6 and 8, the curve of the net charge is often very flat, except for proteins having a His-tag.

Finally, if the protein is only stable within a very narrow range of pH, your choices will be restricted: Go then directly to section 10.4, 10.5, or 10.6 depending on the net charge of the protein at this pH.

10.3.1. Buffers

First of all it must be remembered that buffers are weak acids (or bases) brought near their pKa with their conjugated base (or acid). Their buffer capacity is only efficient in a pH range of:

$$pKa - 0.9 < pH < pKa + 0.9$$

All the protocols given here in the text assume that the crystallizing agents are prepared in water. The pH of the reservoirs is fixed by 50-100 mM buffer. The reservoirs may be prepared by adding 100 μl of a ten-times-concentrated buffer (0.5-1 M) if using neutral salts or organic crystallizing agents (PEGs, alcohol, etc.).

However, when using polyacids or bases (citrate, acetate, phosphate, etc.), their stock solution in water must be adjusted to the right pH, because they are themselves weak acids or bases with their own buffer capacity. Sometimes no additional buffer is then necessary.

In general it is recommended to measure the pH of the reservoir once all constituents (buffer, crystallizing agent, additives) are mixed.

10.3.2. The Initial Test

Prepare drops (or dialysis buttons) at the three pHs to verify the stability of the protein with time. Mix 2 μl of your most concentrated protein solution with 2 μl of buffer. If your protein has free cysteines, remember every 2 or 3 days to add fresh β-mercaptoethanol in the reservoir (or fresh DTT when working with dialysis buttons). See *Reducing Agents* in the *A-Z Tips* section of this book.

10.4. Crystallization of Proteins Having a Positive Net Charge

In 1888, Hofmeister observed that the ability of salts to precipitate hen egg white proteins (the major protein being the negatively charged ovalbumin) follows the series:

> sulfate^{2-} > phosphate^{2-} > acetate^{-} > citrate^{3-} > tartrate^{2-} > bicarbonate^{-} > chromate^{2-} > chloride^{-} > nitrate^{-} >> chlorate^{-}

Sulfate ions are called lyotropic whereas chlorate^{-} (ClO_3^{-}) and thiocyanate (SCN^{-}) are chaotropic.

The results of the solubility measurements of positively charged proteins show an inversion of the Hofmeister anion series, which becomes:

> thiocyanate (SCN^{-}) ~ para-toluene sulfonate (pTS^{-}) > nitrate (NO_3^{-}) > Cl^{-} > acetate ($CH3CO_2^{-}$) ~ phosphate ($H2PO_4^{-}$) > citrate ($HC_8H_5O_6$) $^{2-}$

This can be tested with lysozyme or bovine pancreatic trypsin inhibitor (BPTI).

10.4.1. Range of Salt Concentration to Use

The first step is to find the limit between nucleation and precipitation. Prepare a Linbro box with the following reservoirs:

Salt	stock solution	A	B	C	D
KSCN or Na pTS[*]	1.5 M	0.1 M	0.25 M	0.5 M	1.0 M
NaCl	5.0 M	0.5 M	1.0 M	2.0 M	3.0 M
Na acetate	3.5 M	1.0 M	2.0 M	2.5 M	3.0 M
ammonium phosphate[#]	3.0 M	0.5 M	1.0 M	2.0 M	3.0 M
ammonium citrate[#]	3.5 M	0.5 M	1.0 M	2.0 M	3.0 M
ammonium sulfate	3.5 M	0.5 M	1.0 M	2.0 M	3.0 M

* use KSCN if your protein contains more Arg than Lys. Use Na pTs (para-toluene sulfonate) if the protein contains more Lys than Arg.
\# adjust these stock solutions at the pH of crystallization (see section 10.3).

If you have little protein, prepare 2 ml reservoirs. Use a 5 μl dialysis button with your protein, and transfer it from A to D every 2-3 days until precipitation is observed. Eventually transfer it backwards to verify that the precipitate dissolves again.

If you have more protein, or you need the results quickly, prepare 1 ml reservoirs, and prepare hanging drops with 2 μl of your protein and 2 μl of the corresponding reservoir.

10.4.2. Crystallization

Once the range of salt is found, prepare a new box with the salt concentrations ranging from the last concentration where the drop remained clear to the first concentration where precipitation was observed. Wait one week before drawing conclusions. If crystallization (or precipitation) is observed for × molar NaCl, it usually occurs at about 0.1 × M KSCN and about 2 × M ammonium sulfate.

If no crystal is obtained, or unsatisfactory crystals are grown, the conditions may be optimized by testing salts of the same chemical class:

- sulfonates (p-toluene sulfonate): propane sulfonate, benzoate sulfonate, betaine, or sulfonates usually used as buffer (MOPS, CAPS, MES, etc.)
- halides (NaCl): NaBr, NaI, NaF
- carboxylic acids (acetate, citrate): propionate, tartrate, maleate

For the very soluble snake venom toxins we obtained the best crystals with thiocyanate. The number of crystals could be optimized by playing with the cation (ammonium instead of Na or K).[3]

10.5. Crystallization of Proteins Having a Negative Net Charge

An approach similar to that in section 10.4 is used for negatively charged proteins, except that:

- Since thiocyanate is one of the least efficient anions in the Hofmeister series, it is not tested because it becomes denaturing at high concentrations.

- For proteins containing more acidic residues (Asp and Glu) than basic ones, divalent cations are worth testing. It must be emphasized that the series of anions we have tested is much more diversified than the cations, and we expect to see effects also when a larger variety of cations (tetra-alkylammonium salts, organic bases, metals of higher valency, etc.) are used.

This strategy was developed with collagenase from *Hypoderma lineatum* as a model protein.[4] It was confirmed for Grb2 (growth factor receptor bound protein 2),[5] amylase, and parvalbumin.

10.5.1 Range of Salt Concentration to Use

The first step is to find the limit between nucleation and precipitation. Prepare a Linbro box with the following reservoirs:

Salt	stock solution	A	B	C	D
ammonium sulfate	*3.5 M*	0.5 M	1.0 M	1.5 M	2.0 M
ammonium phosphate[#]	*3.0 M*	0.5 M	1.0 M	1.5 M	2.0 M
ammonium citrate[#]	*3.5 M*	0.5 M	1.0 M	1.5 M	2.0 M
ammonium acetate[#]	*3.5 M*	1.0 M	1.5 M	2.0 M	3.0 M
Mg sulfate	*2.0 M*	0.5 M	1.0 M	1.5 M	2.0 M
Ca chloride	*3.5 M*	0.5 M	1.0 M	1.5 M	2.0 M

[#] adjust these stock solutions at the pH of crystallization (see section 10.3).

If you have little protein, prepare 2 ml reservoirs. Use a 5 μl dialysis button with your protein, and transfer it from A to D every 2-3 days until precipitation is observed. Eventually transfer it backwards to verify that the precipitate dissolves again.

If you have more protein, or you need the results quickly, prepare 1 ml reservoirs, and hanging drops with 2 μl of your protein and 2 μl of the corresponding reservoir.

10.5.2. Crystallization

Once the range of salt concentration is found, prepare a new box with the crystallizing agent concentrations ranging from the last concentration where the drop remained clear to the first concentration where precipitation was observed. Wait a week before drawing conclusions.

If for a given salt no crystal is obtained, or unsatisfactory crystals are grown, the conditions may be optimized by testing salts of the same chemical class:

- halides (NaCl): NaBr, NaI, NaF
- carboxylic acids (acetate, citrate), try propionate, tartrate, maleate

Another option is to test other cations:

- replace ammonium with Li, Na, K
- add 5-10 mM Co, Cu, Ni, Zn, or lanthanides

10.6. Crystallization of Proteins Having a Net Charge of about Zero

At a pH near the pI, the net charge of the macromolecule is near zero. This means the protein is at its lowest solubility compared to a higher net charge. The effect of salts on the solubility of a protein near its pI is illustrated by carboxyhemoglobin.[6] In our hands the effect of ionic strength is poor, and crystallization is more successful with organic crystallizing agents, such as PEGs and MPD. However, it is worth testing salts because they can act by ion binding and the net charge becomes different from zero. For example, if a protein of zero net charge binds some cations, the net charge becomes positive. Therefore, if PEGs and MPD are not successful, test the protocols in sections 10.4 and 10.5 also.

10.6.1. Range of Concentration to Use

The first step is to find the limit between nucleation and precipitation. Prepare a Linbro box with the following reservoirs:

Crystallizing agent*	stock solution	1	2	3	4	5	6
PEG 400	80% (w/v)	2.5%	5%	10%	15%	20%	30%
PEG 4000	50% (w/v)	2.5%	5%	10%	15%	20%	30%
PEG 8000	30% (w/v)	2.5%	5%	10%	15%	20%	30%
MPD	80% (w/v)	2.5%	5%	10%	15%	20%	30%

*It is recommended to add some salt (e.g., 0.2 M NaCl) for ionic strength control.

If you have little protein, prepare 2 ml reservoirs. Use a 5 µl dialysis button with your protein, and transfer it from A to D every 2-3 days until precipitation is observed. Eventually transfer it backwards to verify if the precipitate dissolves again.

If you have more protein, or need the results quickly, prepare the test by the batch method (because vapor diffusion is very slow with PEGs):

- Prepare 1 ml reservoirs.
- Make a series of solutions which are at twice the reservoir concentrations.
- Prepare the drops with 2 µl of your protein and 2 µl of the corresponding twice concentrated reservoir solutions.

10.6.2. Crystallization

Once the range of concentration is found, prepare a new box with the crystallizing agent concentrations ranging from the last concentration where the drop remained clear to the first concentration where precipitation was observed. Wait a week before drawing conclusions.

If no crystal is obtained, or unsatisfactory crystals are grown, the conditions may be optimized by testing mixtures of salts with these organic crystallizing agents. The miscibility of such mixtures can be limited. See appendix A 10.3 of this chapter.

10.7. Optimizing the Nucleation Rate

Once some solid phase appears, the conditions must be refined so that large single crystals grow. The first task is to plot the observations in a schematic phase diagram (Figure 10.1).

10.7.1. The Phase Diagram

10.7.1.1. Solubility Curve

The solubility curve delineates the under- and supersaturated zones. It corresponds to the situation at the end of the process of crystal growth:

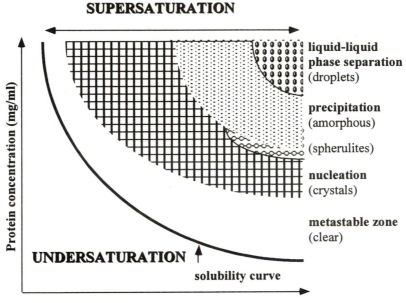

Figure 10.1. Schematic phase diagram showing the different zones and types of solid phase.

The different types of solid phase shown here are rarely encountered in one single set of conditions: The purpose here is to show where each type of solid is located in relation to the others. Note that the precipitation zone can become very close to the solubility curve (not shown) when solubility is very low (right part) or when the solubility curve is very steep (left part).

Additional crystalline protein does not dissolve, but adding reservoir solution without protein leads to the dissolution of the protein crystals. Protein solubility is defined as the concentration of soluble protein, in equilibrium with the crystalline form, at given temperature and pH values, and in the presence of a given concentration of solvent compounds (i.e., water, buffer, crystallizing agents, stabilizers, additives).

Below the solubility curve the solution is undersaturated and the biological macromolecule will never crystallize. Above the solubility curve the concentration of the biological macromolecule is higher than the concentration at equilibrium. This corresponds to the *supersaturation* zone. The protein solution contains an excess of protein which will appear as a solid phase.

The *rate* of supersaturation, is defined as C_p/C_S: The ratio of the biological macromolecule concentration over the solubility value. The higher the supersaturation rate, the faster this solid phase appears.

Much protein and time would have to be invested to accurately define solubility values at equilibrium for any given conditions. However, it is very helpful to know, at least within an order of magnitude, the residual protein concentration in contact with the crystals. This can be done as follows:

- Withdraw 1-2 µl of the drop where crystals have grown for at least a week or two.
- Determine the residual protein concentration in the supernatant. Dilute to the minimal volume required for an OD measurement at 280 nm or use the Bradford method.

Measuring the residual protein concentration gives:

- an estimate of the solubility. At low solubility conditions (≤ 1 mg/ml) growing few and large crystals is difficult to achieve.
- the supersaturation (ratio of initial over final protein concentrations) of the starting conditions.
- the amount of protein (the difference of initial minus final protein concentrations) which is available to grow crystals. The higher the solubility, the more protein is available to grow the crystals, which will be larger if the nucleation rate is reduced.
- the *slope* of the solubility curve, by measuring the residual protein concentration at different values of a given variable, e.g., ionic strength, temperature, pH, etc. This helps for extrapolating to the nucleation zone at lower or higher values of the variable, depending on whether the solubility variation is steep or smooth.
- a guide for the preparation of seeding experiments.
- the amount of protein to include in the solutions used for mounting the crystals. Very often reservoir solution is used to transfer the crystal. In fact, this can be done safely only if the remaining protein concentration in the drop is quite low (≤ 0.5 mg/ml). If the solubility value is high, then the crystal will start to dissolve since the reservoir contains no protein.

10.7.1.2. Metastable Zone

In the metastable zone, the critical supersaturation is not yet reached. Spontaneous nucleation does not occur unless it is induced by vibrations or introduction of a particle that will promote heterogeneous nucleation. The metastable zone is much larger when the protein solubility is very low, in which case it becomes extremely difficult to control the nucleation rate. One can seed crystals into the metastable zone. The crystals will then grow, fed by the amount of protein in supersaturation. When a low solubility system cannot be brought to higher solubility for technical reasons, seeding remains nearly the only way to grow large single crystals.

10.7.1.3. Nucleation Zone

In the nucleation zone, crystals appear spontaneously. The nucleation rate, defined as the number of nuclei formed per unit volume and unit time, is linked to:
- supersaturation: when increasing the supersaturation, the number of crystals increases and they appear earlier.
- the number of molecules per unit volume. When the solute is sparingly soluble, the solution remains in a metastable state over long periods. Nucleation requires much higher supersaturation, and the nucleation rates become drastic.

If spherulites are observed, then the conditions are located at the upper border of the nucleation zone, most often at high concentrations of crystallizing agent. The zone of spherulites shown in Figure 10.1 may extend higher along the precipitation zone.

10.7.1.4. Precipitation Zone

Precipitation occurs at very high supersaturation (for example, 30 to 100 times the solubility value in the case of lysozyme). Insoluble protein rapidly separates from the solution into an amorphous state. If the solution is centrifuged, the supernatant is in fact still supersaturated and crystallization may occur. To differentiate amorphous precipitate from microcrystals, fresh drops can be seeded with this "precipitate": Precipitate dissolves whereas microcrystals grow.

When liquid-liquid phase separation is observed (droplets in the protein solution), the conditions are located in the upper part of the precipitation zone, with high concentrations of crystallizing agent. To reach the nucleation zone, the concentration of the crystallizing agent should be reduced, and possibly the protein concentration as well.

10.7.2. Optimal Conditions at Moderate Concentrations of the Crystallizing Agent

Once a solid phase is obtained, the conditions must be optimized to grow large single crystals. The schematic phase diagram is like a map, a guide to the nucleation zone where only few crystals grow. In our hands the best zone is at moderate salt concentration and high protein concentration, because:

- There is more protein to feed the crystals, so they will be larger. A large single crystal needs at least 10-50 µg of protein.
- The nucleation rate is easier to control.
- The effects of pH and of temperature are amplified. As explained in section 10.3, the higher the net charge, the higher the solubility. With respect to temperature, the solubility may be direct (increasing with the temperature) or retrograde (decreasing with the temperature). This is due to the nature of the protein salt: BPTI in NaCl or ammonium sulfate has a retrograde behavior, whereas BPTI in KSCN has a direct solubility.

Acknowledgement

Marie-Thérèse Latreille is acknowledged for her technical help in determining the miscibility curves.

References and Further Reading

1. Riès-Kautt M, Ducruix A: **Inferences from physico-chemical studies of crystallogenesis and the precrystalline stage.** *Methods Enzymol* 1997, **276**:23-59.

Theoretical background of the physico-chemical approach, especially the effect of ions in solutions and on protein crystallization.

2. Riès-Kautt M, Ducruix A: **From solution to crystals with a physico-chemical aspect.** In *Crystallization of Nucleic Acids and Proteins: A Practical Approach.* Edited by Ducruix A and Giegé R. IRL/Oxford Press; in press.

Theoretical background of our approach.

3. Ménez R, Ducruix A: **A toxin that recognizes muscarinic acetylcholine receptors.** *J Mol Biol* 1993, **232**:997-998.

An example of crystallization in a sodium salt, whereas the previous analogs (reference cited in the paper) were crystallized in the potassium salt.

4. Carbonnaux C, Riès-Kautt M, Ducruix A: **Relative effectiveness of various anions on the solubility of acidic *Hypoderma lineatum* collagenase at pH 7.2.** *Protein Science* 1995, 4:2123-2128.

Quantification of the effect of anions on the solubility of an acidic protein.

5. Guilloteau JP, Fromage N, Riès-Kautt M, Reboul S, Bocquet D, Dubois H, Faucher D, Colonna C, Ducruix A, Becquart J: **Solubilization, stabilization and crystallization of a modular protein Grb2.** *Proteins: Structure, Function, and Genetics* 1996, **25**:112-119

Illustration of the effect of anions on the crystallization of an acidic protein.

6. Green AA: **The solubility of hemoglobin in solutions of chlorides and sulfates of varying concentrations.** *J Biol Chem* 1932, **95**:47-66.

Quantification of the effect of pH on the solubility of carboxyhemoglobin, an acidic protein.

A 10.1. Spread Sheet for the Calculation of the Protein Net Charge versus pH

pH	α-COOH n=□	Asp + Glu n=□	Tyr n=□	free Cys-SH n=□	α-NH₂ n=□	His n=□	Lys n=□	Arg n=□	Zp
2.00	- 0.03 n =				+ 1 n =	+ 1 n =	+ 1 n =	+ 1 n =	
2.50	- 0.09 n =	- 0.01 n =			+ 1 n =	+ 1 n =	+ 1 n =	+ 1 n =	
3.00	- 0.24 n =	- 0.03 n =			+ 1 n =	+ 1 n =	+ 1 n =	+ 1 n =	
3.50	- 0.50 n =	- 0.09 n =			+ 1 n =	+ 1 n =	+ 1 n =	+ 1 n =	
4.00	- 0.76 n =	- 0.24 n =			+ 1 n =	+ 1 n =	+ 1 n =	+ 1 n =	
4.50	- 0.91 n =	- 0.50 n =			+ 1 n =	+ 0.99 n =	+ 1 n =	+ 1 n =	
5.00	- 0.97 n =	- 0.76 n =			+ 1 n =	+ 0.97 n =	+ 1 n =	+ 1 n =	
5.50	- 0.99 n =	- 0.91 n =			+ 0.99 n =	+ 0.91 n =	+ 1 n =	+ 1 n =	
6.00	- 1 n =	- 0.97 n =			+ 0.97 n =	+ 0.76 n =	+ 1 n =	+ 1 n =	
6.50	- 1 n =	- 0.99 n =			+ 0.91 n =	+ 0.50 n =	+ 1 n =	+ 1 n =	
7.00	- 1 n =	- 1 n =			+ 0.76 n =	+ 0.24 n =	+ 1 n =	+ 1 n =	
7.50	- 1 n =	- 1 n =	- 0.01 n =	- 0.02 n =	+ 0.50 n =	+ 0.09 n =	+ 1 n =	+ 1 n =	
8.00	- 1 n =	- 1 n =	- 0.03 n =	- 0.05 n =	+ 0.24 n =	+ 0.03 n =	+ 1 n =	+ 1 n =	
8.50	- 1 n =	- 1 n =	- 0.09 n =	- 0.14 n =	+ 0.09 n =	+ 0.01 n =	+ 0.99 n =	+ 1 n =	
9.00	- 1 n =	- 1 n =	- 0.24 n =	- 0.33 n =	+ 0.03 n =		+ 0.97 n =	+ 1 n =	
9.50	- 1 n =	- 1 n =	- 0.50 n =	- 0.61 n =	+ 0.01 n =		+ 0.91 n =	+ 1 n =	
10.0	- 1 n =	- 1 n =	- 0.76 n =	- 0.83 n =			+ 0.76 n =	+ 0.99 n =	
10.5	- 1 n =	- 1 n =	- 0.91 n =	- 0.94 n =			+ 0.50 n =	+ 0.97 n =	
11.0	- 1 n =	- 1 n =	- 0.97 n =	- 0.98 n =			+ 0.24 n =	+ 0.91 n =	
11.5	- 1 n =	- 1 n =	- 0.99 n =	- 0.99 n =			+ 0.09 n =	+ 0.76 n =	
12.0	- 1 n =	- 1 n =	- 1 n =	- 1 n =			+ 0.03 n =	+ 0.50 n =	
12.5	- 1 n =	- 1 n =	- 1 n =	- 1 n =			+ 0.01 n =	+ 0.24 n =	
13.0	- 1 n =	- 1 n =	- 1 n =	- 1 n =				+ 0.09 n =	

A 10.2. Equations to Calculate a Protein Net Charge

The calculations are performed according to:

$$Z_p = \overset{\alpha NH_2,His,Lys,Arg}{\underset{(+)}{\sum n_{i(+)}}} \times ([H^+]/(K_{a(+)} + [H^+])) - \overset{\alpha COOH,Asp,Glu,Tyr,CysH}{\underset{(-)}{\sum n_{i(-)}}} \times (K_{a(-)}/(K_{a(-)} + [H^+]))$$

with:

$[H^+] = 10^{-pH}$; $K_a = 10^{-pKa}$ for the individual charged groups;

n_i = the number of each type of charged residue.

Meaning	Column code and equation
pH values (from 2 to 12, by increments of 0.5)	c0
[H+]	c1=10^-c0
Charge of one COOH$_{term}$ residue	c2=3.16*10^-4/((3.16*10^-4)+c1)
Charge of one Asp or Glu residue	c3=3.16*10^-5/((3.16*10^-5)+c1)
Charge of one Tyr residue	c4=3.16*10^-10/((3.16*10^-10)+c1)
Charge of one free Cys residue	c5=5.01*10^-10/((5.01*10^-10)+c1)
Charge of one NH$_{2\ term}$ residue	c6=c1/((2.51*10^-8)+c1)
Charge of one His residue	c7=c1/((6.31*10^-7)+c1)
Charge of one Lys residue	c8=c1/((3.98*10^-11)+c1)
Charge of one Arg residue	c9=c1/((10^-12)+c1)
Sum of all positive charges*	c10=(n$_{NH2\ term}$*c6)+(n$_{His}$*c7)+(n$_{Lys}$*c8)+(n$_{Arg}$*c9)
Sum of all negative charges*	c11=(n$_{COOHterm}$*c2)+(n$_{Asp+Glu}$*c3)+(n$_{Tyr}$*c4)+(n$_{Cys}$*c5)
Net charge	c12=c10-c11

*Replace each n_i by the number of the corresponding number of residues.

Example for lysozyme:

c10=(1*c6)+(1*c7)+(6*c8)+(11*c9)

c11=(1*c2)+(9*c3)+(3*c4)+(0*c5)

A 10.3. Miscibility Curves for Organic Crystallizing Agents and Various Salts

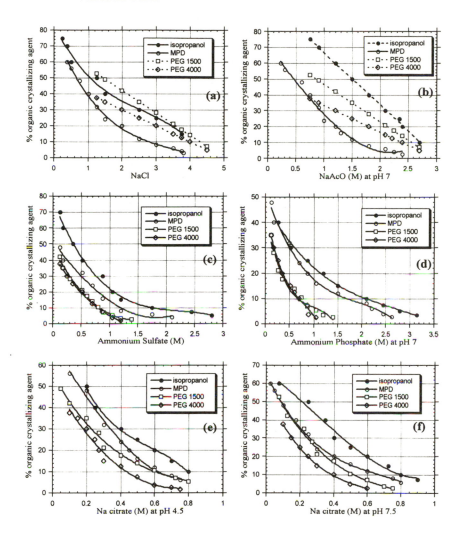

The curves represent the upper limit of miscibility of the indicated compounds. The curves were obtained by mixing various volumes of stock solutions (5.0 M NaCl, 3.5 M NaAcO, 3.5 M ammonium sufate, 3.5 M ammonium phosphate, 1.0 M sodium citrate, 100% (v/v) isopropanol, 80% (v/v) MPD, 70 % (w/v) PEG 1500, 50% (w/v) PEG 4000) to a final volume of 1 ml with water. If larger volumes were mixed, phase separation was observed immediately or within a few hours at $22 \pm 3°C$.

The dotted curves are the maximal concentrations achievable with the stock solutions (without dilution with water) and which present no limit to miscibility. If working with more concentrated stock solutions, miscibility should be verified.

Similarly no phase separation could be reached by mixing the following stock solutions:

- glycerol (60 % v/v) with 5.0 M NaCl, 3.5 M ammonium sufate, 3.5 M ammonium phosphate or 1.0 M sodium citrate.

- NaSCN (1.5 M) or p-toluene sulfonate (1.5 M) with 100% (v/v) isopropanol, 80% (v/v) MPD, 70 % (w/v) PEG 1500, 50% (w/v) PEG 4000 or glycerol (60% v/v).

11

Strategy 3: Reverse Screening

Enrico A. Stura

11

Strategy 3: Reverse Screening

Enrico A. Stura

DIEP, CEA Saclay, France

Reverse screening[1] is to crystallization what differentiation is to a mathematical function: It looks at small changes and analyzes their effect. Hence it is an ideal method for improving crystallization conditions, but it is also useful in screening additives which may be the key to nucleation. Two main principles of reverse screening are:

1. Maximize the degree of supersaturation while screening.

2. Select a non-specific precipitation solution and test various additives. Repeat the tests at various protein conditions.

When screening for initial crystallization conditions, begin with rule 1. If some crystals have already been obtained, start with rule 2 for optimization of the conditions. The laboratory exercises in this chapter use lysozyme, a protein whose crystallization conditions are already known. Therefore we will begin in reverse order, i.e., with rule 2.

11.1. Specificity of Precipitants and Buffers

Rule 2: **A precipitant should induce supersaturation and nothing else; buffers control pH and nothing else.** Precipitants should be replaceable by other precipitants, and buffers by other buffers at the same pH. Buffers and precipitants that follow this condition are considered to be non-specific and allow the control of supersaturation and pH independently of other parameters.

This may be best explained with an example. In the crystallization of lysozyme, NaCl is used as a precipitant. Does the NaCl have a specific effect, or can it be replaced by polyethylene glycol or ammonium sulfate? We will answer this question in the experiment below.

Experiment 11.1. Monomethyl PEG (MPEG) crystallization of lysozyme

Purpose: The application of rule 2 of reverse screening to the crystallization of lysozyme.

This experiment is designed to be carried out at room temperature (17-24° C), on a siliconized cover slip under the microscope. The drop will not dry out for the duration of the experiment. After the experiment, the crystals can be stored as a vapor diffusion setup by adding 0.5 ml of the same precipitant and 0.5 ml of the same buffer to the reservoir.

1. Weigh 10 mg of lyophilized hen egg white lysozyme in a microcentrifuge tube and add 100 µl of a buffer consisting of 50 mM sodium acetate, pH 4.5, to obtain a 100 mg/ml solution of lysozyme. Centrifuge to remove bubbles.

2. Make a stock solution of 50% (w/w) monomethyl polyethylene glycol (MPEG) 5000. It will be necessary to microwave the solution to help it dissolve. Prepare a precipitant solution composed of 30% MPEG, 1 M NaCl, and 50 mM sodium acetate, pH 4.5. Do this while the 50% solution is well mixed but still warm, so that the viscosity will be less of a problem.

3. Mix an equal volume of the lysozyme solution with the precipitant under a microscope. Mix the drop well, for 10-20 seconds. You should be able to see the crystals nucleate within 5-15 minutes. Make a note of the time it takes for the first visible crystals to appear.

4. Repeat the experiment with 500 mM NaCl instead of 1 M. How long do you have to wait for nucleation? Now, instead of using 30% MPEG, use only 20%. Write down your observations.

MPEG and NaCl act as co-precipitants, yet changing the concentration of MPEG is not the same as changing the concentration of NaCl. At this pH, Cl^- forms a salt with lysozyme and reduces its solubility.

5. Try 300 mM $NaNO_3$ (sodium nitrate) instead of 1 M NaCl. Continue on your own to analyze the effect of other salts (suggestions: 150 mM KSCN, 6 mM sodium citrate or vanadate).

   While chloride, nitrate, and thiocyanate are monovalent, citrate and vanadate are polyvalent and more potent additives (co-precipitants).

6. Repeat the experiment, replacing MPEG 5000 with polyethylene glycol (PEG) of other molecular weights. As you decrease the molecular weight of the PEG or MPEG, increase their concentration.

   The effect of changing PEG is not as drastic as that of changing the salt. Unlike the salts, PEG is a non-specific precipitant.

7. Prepare a new precipitant solution consisting of 30% MPEG 5000, and 100 mM imidazole malate, pH 4.5. This new solution omits the NaCl and replaces the sodium acetate with a different buffer of the same pH. Repeat the experiment and streak seed the drop from crystals obtained previously (see Chapter 14, *Seeding*). Substitute other buffers such as aspartate (aspartic acid) or citrate.

   Not all buffers are equal—buffers are salts, too.

8. Dilute the lysozyme solution with 50mM sodium acetate buffer to 50, 25, 10, and 5 mg/ml and repeat the experiment. Try new combinations and different pH levels.

In this experiment you have investigated the "specific effects" of precipitants and buffers, and of changing the protein concentration.

11.2. Search at a High Degree of Supersaturation

In Experiment 11.1 we started with a set of crystallization conditions and analyzed the role of each component of a precipitant solution. This is the "reverse" of most screening methods, which sample many different precipitant solutions to find one that will produce crystals. Reverse screening takes a more systematic approach by analyzing the effect of changes in the precipitant composition in relation to the amount and morphology of the resultant precipitates.

Rule 1: **Since the probability of nucleation is proportional to the number of molecules in the supersaturation state, screening should be done where the gradient of the solubility curve (phase diagram) is steep, as this maximizes the chances of getting crystals.**

In practice, such conditions are identified by a large change in precipitation when the precipitant concentration is increased by a small amount. Reverse screening selects a salt-based and a PEG-based precipitant and proceeds as described in Experiment 11.1, looking at changes in precipitation until crystals are obtained.

11.3. Screening

The first step in screening is to find the threshold at which the protein starts to precipitate, because crystals or microcrystals are obtained close to this point. This is done by either increasing the protein concentration or adding a precipitating agent or both. Even after some of the protein has precipitated, the solution remains supersaturated and nucleation and growth remain possible under such conditions.

It is important to try to identify as many crystal forms as possible irrespective of their size and quality. The more different forms that are identified, the more likely it is that one of them will lead to diffraction quality crystals. Size and good diffraction are not related. Many large crystals diffract poorly, and high-resolution structures have been obtained from small crystals. The choice of screening strategy depends on the amount of protein available. The methods discussed here aim to minimize materials and setup time, although results may be obtained more slowly.

11.4. Solubility Evaluation

There is great variation in protein solubility, and it is wasteful to test all the conditions in a standard screen. Unless the protein is unstable because of proteases, oxidation, or non-specific aggregation, experiments can be postponed until results from previous experiments have been obtained. The ability of PEG to give fast results, as demonstrated in Experiment 11.1, can be utilized to evaluate the protein solubility.

Experiment 11.2. Solubility screening

Purpose: To apply the principles of reverse screening to an experimental protein.

In this experiment use vapor diffusion sitting (preferable) or hanging drops. You can use an experimental protein, as by now you should have a good idea of what results you obtain with lysozyme.

1. Make your own set of PEG solutions to cover a wide range of concentrations, pH values, and different molecular weights. Alternatively, use the footprint #1 screen (see reference,[2] http://bmbsgi13.leeds.ac.uk.wwwprg/stura/cryst/screen.html) or the PEG screen in Table 11.1.

2. Start with a solution of approximately 20% PEG 4000, with buffer at pH ~7.0 (footprint #1, condition 2C). Use this as the reservoir solution. Place 1 μl of your protein solution on the inverted pot,[3] microbridge, or siliconized cover slip. Add 1 μl reservoir solution to the protein drop. Observe under the microscope for 20 seconds, then stir the drop. To evaluate the precipitates, use the results from the lysozyme Experiment 11.1 with $NaNO_3$ as a positive comparison and those obtained with vanadate as a negative comparison.

   If the drop remains clear, set up successive drops each at 5% higher PEG concentrations. If the PEG concentration needed to precipitate the protein is higher than 30%, it is best to repeat the screen with protein at double the concentration. Above 30% PEG

4000 or 80% PEG 400 it becomes possible to crystallize some of the components of the buffer system that the protein is in, plus the viscosity of the solution leads to measuring and mixing inaccuracies.

3. Set up further drops with PEG at different pH levels to evaluate the change in solubility with pH. Most proteins are less soluble at acidic pH than at basic pH, and less soluble around their isoelectric point (pI). Set up some drops with pH values one unit above and one unit below the protein's pI.

4. While solubility screening with PEG is virtually instantaneous, setups with salts will require equilibration for at least 8 hours. Set up drops with 0.083 M ammonium sulfate or phosphate or 0.05 M sodium citrate for each 1% PEG 4000 at different pH. This is just a rough guideline and many proteins deviate from this.

11.5. Additive Screening

During the initial solubility test, the concentration of precipitant needed to induce supersaturation has been evaluated. Subsequently, as in Experiment 11.1, buffers, precipitants, and salts are now systematically replaced to define their roles and additives are introduced (Table 11.2). In order to properly evaluate the effect of the additives, it is important **not** to make unwanted changes in the other parameters.

Step 1. Make a "working solution."
Begin by mixing a "working solution" (ws1) of 15-200 ml, based on the precipitant concentrations and buffer determined in the initial solubility test. Preparing a large volume will enhance the reproducibility and give greater significance to the changes that result from an additive screen. In addition, advance preparation of the reservoirs will be faster than mixing a new reservoir each time.

Step 2. Screen the additives.
The effects of additives can now be screened. For example if the ws1 is 16% PEG 6000, 100 mM imidazole malate, and ethanol is chosen as the

additive, prepare the following reservoirs: ˙

- 95% ws1 + 5% ethanol
- 95% ws1 + 2% ethanol + 3% water
- 95% ws1 + 1% ethanol + 4% water
- and so on.

When protein is at a premium, do just the 5% setup on day 1. Do the others only if a change is observed.

If the additive results in an improvement in the drop results, it is used as the basis for a new formulation of the working solution (ws2). In the example above, ws2 might be 12% PEG 6000, 100 mM imidazole malate, and 5% ethanol. The concentration of the precipitant is reduced from 16% in ws1 to 12% in ws2 because of the contributory effects of the ethanol.

Continue screening additives in ws1 to make ws3, ws4, etc. Do not assume that a single additive will suffice.

- If the additives are not related, add the new additive to ws2. Later, add the optimized amount to ws1 to check if the second additive alone suffices.

- If the additives are related, add the new one to both ws1 and ws2. The first looks at the effect of additive 2 as a replacement for additive 1. The second looks at its cumulative effect with additive 1.

11.5.1. Use of Additives

Table 11.2 lists commonly used additives, grouped according to their usage, with suggested concentration ranges. For each protein, it is suggested that one additive from each group, as well as combinations between groups, be tested.

Initially an amount within the range for the compound is added to the working solution. If a large change is observed the additive concentration is reduced two- to ten-fold. When no change is observed, a higher concentration may be tried, although it is better to try a different compound.

1.5.2. Magic Solutions

As the "working solution" is further refined and can reliably yield crystals for X-ray data collection, just like magic, it is referred to as the "magic solution." These are the conditions published together with the structure. The magic solution can be used in searching for new crystal forms, heavy atom soaking, etc.

11.6. Heavy Atom Soaking and Cryo-Crystallography

In the method of multiple isomorphous replacement to derive the phases for X-ray structure determination, heavy atom compound soaking is often used. The success of such experiments depends on the solubility changes that the heavy atom compounds induce. In Experiment 11.3 below, the magic solution is used to check for solubility changes that may occur when the heavy atom compound is added. The heavy atom compound is treated as an additive in the crystallization. Solubility changes are compensated for by making changes in the soak solution. A cryo-solvent can be also considered an additive and treated in the same manner. Cryo-solvents and heavy atom soaking can also be combined into a single soak solution. Since it is difficult to soak heavy atom compounds in citrate solutions, replace these by another salt or a PEG-based precipitant as suggested in Experiment 11.3.

 ### Experiment 11.3: Solubility screening applied to heavy atom soaking

Purpose: To determine a soak solution for introducing heavy atom compounds into a protein crystallized with citrate. (The protocol can be adjusted as appropriate for other crystallization conditions as well.)

1. Replace the citrate with a 50-100 mM succinate buffer at the same pH and MPEG 5000 (1.5 M citrate = approximately 30% MPEG

5000). Introduce your crystal into this solution. If the crystal dissolves, add NaCl or other salts to increase the ionic strength (see Table 2), then try with a new crystal. Add more salt or change salts until the crystal is stable.

2. Add 0.5-10 mM heavy atom solution to the magic solution or the solution determined in step 1 and set up a crystallization trial drop. If there is an increase in precipitation, the heavy atom has passed the solubility test. If not, follow step 2a. If the change in the amount of precipitation is great, follow step 2b.

2a. Increase the precipitant concentration of the magic solution by 10%. (Example: If the magic solution contained 16% PEG 4000, the new concentration would be 17.5%.) If the crystals are still cracking or dissolving, make either NaCl or MPEG additions as described in step 1 as many times as is reasonable to increase the stabilization power of the solution.

2b. Decrease by 2 or 3 times the concentration of heavy atom compound until the amount of precipitate becomes reasonable. When soaking crystals with the reduced heavy atom solution, use a volume as large as needed to provide 1-2 molar equivalents of the heavy atom for the macromolecules in the crystal(s).

11.6.1. Some Additional Tips In Preparing Heavy Atom Derivatives

1. The data collected for the derivatized crystal may not be isomorphous with the native. If there is no substantial loss in the diffraction resolution, the lack of isomorphism may be due to a specific effect. Try collecting a new native data set in the presence of a "light atom" compound similar to the one used in the derivative, e.g., zinc instead of mercury, magnesium instead of samarium, etc. You can test whether the same change in lattice parameters may be induced by organic solvents, alcohols, etc. If the same lattice parameter changes that occur with heavy atom soaking can be achieved with the organic solvent, a matched set of native-

derivative data can be used for phasing even when the collected heavy atom data set is not isomorphous.

2. Derivatization with heavy metals tends to be more successful around neutral pH. Even if the crystals grow best at a basic or acidic pH, they may be stable enough at neutral pH levels to be transferred if compensating changes in the precipitant concentration are made.

11.7. Summary

Either one of the steps in reverse screening can be applied at any time. They serve to analyze the importance of any parameter as a function of solubility or, in conjunction with seeding, of nucleation and growth. Reverse screening is not the only strategy for crystallization, but its two rules are easy to use. To go forward, screen in reverse.

References and Further Reading

1. Stura EA, Satterthwait AC, Calvo JC, Kaslow DC, Wilson IA: **Reverse screening.** *Acta Cryst* 1994, **D50**:448-455.

   Introduction to theoretical and practical considerations in reverse screening.

2. Stura EA, Nemerow GR, Wilson IA: **Strategies in the crystallization of glycoproteins and protein complexes.** *J Cryst Growth* 1992, **122**:273-285.

   Footprint screen #1. Heterogeneity as in glycoproteins and protein complexes.

3. Stura EA, Johnson DL, Inglese J, Smith JM, Benkovic SJ, Wilson IA: **Preliminary crystallographic investigations of glycinamide ribonucleotide phosphorylase.** *J Biol Chem* 1989, **264**:9703-9706.

   Use of sitting drop vapor diffusion pedestals.

4. Stura EA, Wilson IA: **Application of the streak seeding technique in protein crystallization.** *J Cryst Growth* 1991, **110**:270-282.

   Introduction to streak seeding and its analytical uses as well as its use for propagation of crystals.

5. Stura EA, Ruf W, Wilson IA: **Crystallization and preliminary crystallo-graphic data for a ternary complex between tissue factor, factor VIIa and a BPTI-derived inhibitor.** *J Cryst Growth* 1996, **168**:260-269.

   First published case of an epitaxial jump. Crystallization screening using 1.5 mg of protein. Calcium and NaCl to modulate protein solubility.

A 11.1. Tables for a PEG Screen and Common Crystallization Additives

Table 11.1. PEG screen.

PEG Type	Monomethyl ether PEG				Polyethylene glycol (PEG)		
PEG Type M.W.	MPEG 550	MPEG 2000	MPEG 5000		PEG 600	PEG 4000	PEG 10000
Buffer	100 mM HEPES	100 mM sodium cacodylate	100 mM sodium acetate		100 mM HEPES	200 mM imidazole malate	100 mM ammonium acetate
pH	pH 8.2	pH 6.5	pH 5.5		pH 7.5	pH 6.0	pH 4.5
position*	1	3	5		2	4	6
				A	18%	8%	9%
A	*30%*	*18%*	*12%*	*B*	*27%*	*15%*	*15%*
B	40%	27%	18%	C	36%	20%	22.5%
C	50%	36%	24%	D	45%	30%	27%
D	60%	45%	36%				

*Position numbers match with crystallization website at:
http://bmbsgi13.leeds.ac.uk.wwwprg/stura/cryst/screen.html.

This screen is set up completely only if the protein supply is not at a premium. Otherwise, the "average" protein would respond well to row *1-5A, 2-6B* (shaded row, shown in italics). Column 4 is designed as a comparison, if you have done footprint screen #1 on your protein.[2] The buffers in the other columns have been chosen to maximize compatibility with the use of divalent metals (Table 11.2). The even-numbered columns (PEG; right part of Table 11.1) are offset from those of the odd rows (monomethyl ether PEG; left part of the table) so that row A of the odd columns matches in precipitation strength with the even columns. This enables the scanning of a wider precipitant range than with rows of equal strengths.

Table 11.2. Commonly used additives in the crystallization of biological macromolecules.

Alcohols		Monovalent salts		Phosphate mimics	
methanol	5-15%	NaCl	25mM-5M	sulfate	.05-4.0 M
ethanol	5-20%	KCl	25mM-4M	vanadate	1-20 mM
2-propanol	5-20%	KSCN	10-500 mM	molybdate	1-20 mM
Glycols		LiCl	25mM-5M	tungstate	1-50 mM
Ethylene glycol	15-45%	**Divalent metals**		cacodylate	25-200 mM
PEG 200-600	5-50%	$MgCl_2/MgSO_4$	5-100 mM	pyrophosphate	10-100 mM
PEG 20000	1-5%	$CaCl_2$/Ca acetate	1-20 mM	**Trivalent metals**	
Diols		$NiCl_2/NiSO_4$	0.5-10 mM	Lanthanide Cl_3	.005-0.3%
MPD	0.5-55%	$MnCl_2/$	1-100 mM	**Polyvalent salts**	
		$MnSO_4$			
1,6-hexanediol	0.5-60%	$ZnCl_2/ZnSO_4$	0.2-5 mM	$(NH_3)_2SO_4$	10-500 mM
2,5-hexanediol	0.5-60%	$CdCl_2/CdSO_4$	0.1-2 mM	Citrate	0.1-1.5M
glycerol and sugars		$CuCl_2/CuSO_4$	0.2-5 mM	Succinate	0.1-1M
glycerol	5-45%	Fe(II) citrate	0.1-10 mM	**Diamines**	
xylitol	5-35%	$CoCl_2/CoSO_4$	1-100 mM	spermine	0.5-10 mM
mannitol	5-35%	**Reducing agents**		spermidine	1-10 mM
sorbitol	5-35%	β-mercapto-ethanol[&]	0.2-10 mM	**Protease inhibitors**	
trehalose	3-35%	dithiothreitol[&]	0.1-5 mM	PMSF*	1-10 mM
glucose	10-35%	glutathione[&]	0.1-5 mM	benzamidine*	1-50 mM
sucrose	10-35%	**Solubilization agents**		**Chelating agents**	
Detergents/amphiphiles		dioxane	2-20%	EDTA	1-10 mM
β-octyl glucoside*	0.1-1%	DMSO	2-20%	EGTA	1-10 mM
heptane-1,2,3-triol*	1-10%	hexafluoro-propanol	1-5%	Imidazole	20-200mM

* Typically added to macromolecular solution and not to reservoir.

& Typically added to macromolecular solution and also to reservoir.

Abbreviations: MPD: 4-methyl-2,4-pentane diol; PMSF: phenyl methyl sulfonyl fluoride.

In addition to the above, nucleotides are commonly used as additives for proteins that bind them and likewise for the dinucleotides: NAD, NADH, NADP, NADPH. Sodium azide is not included in this list but is seldom omitted from crystallization solutions, even if it is rarely mentioned in crystallization reports.

12

Strategy 4: Imperial College Grid Screen

Lesley Lloyd Haire

12

Strategy 4: Imperial College Grid Screen

Lesley Lloyd Haire

*National Institute for Medical Research,
London, U.K.*

The Imperial College Screen was designed by P. Brick and L. Lloyd Haire[1] as a simple systematic grid screen for new proteins, to give as much information as possible on the protein solubility under various conditions. The screen consists of 60 premixed solutions. The two major protein precipitants, PEG and ammonium sulfate, are screened at four pH values and five concentrations. We decided to use the other precipitants, PEG 4000, ammonium sulfate, phosphate, citrate, isopropanol, MPD, PEG 4000/NaCl, and PEG 4000/LiCl, at just one pH, 6.8, to conserve protein.

This screen differs from the Hampton Research grid screens in that the latter include more pH values and much coarser intervals. Other precipitants are included in their sparse matrix screens, but in combination with additives, which makes interpretation of solubility behavior more difficult.

12.1. The Imperial College Screen

Buffers
Sodium acetate pH 4.6
MES pH 6.1
PIPES pH 6.8
Tris HCl pH 8.3

Precipitants	Concentration
Ammonium sulfate	1.1, 1.4, 1.7, 2.0, 2.3M
PEG 4000	5, 10, 15, 25%
Sodium potassium phosphate	1.1, 1.4, 1.7, 2.0, 2.3M
Sodium citrate pH 6.4	1.2, 1.4, 1.6M
MPD	10, 20, 30, 40, 50%
Isopropanol	5, 10, 15, 20, 25%
PEG 4000 + NaCl	5, 10, 20% + 1M NaCl
PEG 4000 + LiCl	5, 10, 20% + 1M LiCl

Two µl hanging drops (final volume) or microbatch drops are used to set up the experiments, which require approximately 0.6 mg protein (assuming a protein stock solution of 10 mg/ml). It is advisable to set up the drops containing isopropanol or MPD as vapor diffusion, since these volatile solvents tend to dissolve in the paraffin oil used in the microbatch technique. This screen has yielded crystals subsequently used for structure elucidation such as lysU[2] and firefly luciferase.3[2]

12.2. The Imperial College Supplemental Screen

A supplementary additives screen was devised by N. Chayen and L. Lloyd. This screen may be used during optimization after preliminary screening with PEG or ammonium sulfate at pH ~7. The screen consists of 11 solutions of ammonium sulfate at 1.7 M in 100 mM PIPES, pH 6.8, and 11 solutions of PEG 4000 at 20% (w/v) in 100 mM PIPES, pH 6.8, containing the following additives at the given concentrations:

Additives

1. 10 mM phenol
2. 10 mM spermidine
3. 10 mM spermine
4. 10 mM CsCl
5. 10 mM cobalt hexamine chloride
6. 1% (w/v) β-octyl glucoside
7. 50 mM malonate

8. 50 mM maleate
9. 50 mM ethylene diamine
10. 50 mM alanine
11. 10 mM thymol

These solutions may be used with either the microbatch or hanging drop methods.

References and Further Reading

1. Lloyd LF: Ph.D. Thesis *University of London*, 1996.

   Thesis on "strategies for protein crystal growth-screening and optimization" giving details of the Imperial College Screen.

2 Onesti S, Theoclitou ME, Wittung L, Miller AD, Plateau P, Blanquet S, Brick P, Pernilla E: **Crystallization and preliminary diffraction studies of *Escherichia coli* lysyl-tRNA synthetase (LysU).** *J Mol Biol* 1994, **243**:123-125.

   Successful crystallization of an enzyme with the Imperial College Screen and the hanging drop vapor diffusion method.

3. Conti E, Lloyd LF, Akins J, Franks NP, Brick P: **Crystallization and preliminary diffraction studies of firefly luciferase from *Photinus pyralis*.** *Acta Cryst* 1996, **D52 (4)**:876-878.

   An example of microbatch crystallization with the Imperial College Screen.

13

Interpretation of the Crystallization Drop Results

Johan Philip Zeelen

13

Interpretation of the Crystallization Drop Results

Johan Philip Zeelen

*Max Planck Institute of Biophysics,
Frankfurt am Main, Germany*

13.1. Result Interpretation and Type of Screen

The major difficulty in the crystallization of proteins is the identification of conditions that, after optimization, will produce crystals suitable for X-ray diffraction analysis. Since good quality crystals are not usually obtained in the first screen, it is necessary to get leads from whatever results that are produced. The approach to "reading" the crystallization drop will depend somewhat on the type of screen used.

The **sparse matrix** screen is used for the identification of those areas in variable space that have some chance of producing crystals. The crystallization experiments are examined with a stereomicroscope 1) immediately after setup, 2) each day for the first week, and 3) once a week for several weeks. When storage space is available the experiment may be continued for as long as 2 years. The conditions that cause precipitation or crystal growth are identified as is the rate at which they occur, because the rate of formation is important for optimization of the precipitant concentration (see Chapter 9, *A Flexible Sparse Matrix Screen*, section 3).

The advantage of the sparse matrix screen is that it tests many parameters with a limited amount of protein. This might result in the identification of different crystal forms, but often only information about the solubility of the protein is obtained. Therefore, careful examination of the results is required in order to identify promising leads.

In the **grid screen** we screen every precipitant type in separate trays, varying pH (4.5-9.5 in 1.0 pH intervals) and precipitant concentration (4 levels). The results of a grid screen can be plotted in a phase diagram. The phase diagram shows the different zones for soluble protein (Score=0), denaturation (Score=3), precipitation (Score=3-4), and nucleation (Score=6-9). Extreme pH values, where the protein becomes denatured, should be avoided.

The grid screen (pH vs. precipitant) requires greater amounts of protein but is easier for the inexperienced crystallizer to interpret. Unlike the sparse matrix screen, the grid screen is systematic and the trends in precipitation behavior are easy to identify or infer.

In the final stages of optimization, when we are aiming for highly ordered, big crystals, it is better not to disturb the crystal growth process, so wait one or two weeks before checking the experiment.

13.1.1 The Stereomicroscope

For the examination of the crystallization conditions we use a stereomicroscope. The stereomicroscope should have an observation platform that is smooth and big enough to support the tray when looking at all the drops. For transmitted light use a separate light source and a fiber-optic light guide with bright and dark field illumination. If the only microscope available is one with the light source in the observation platform, place a Perspex (plexiglass) plate between the platform and the tray, and work quickly to prevent heating of the drop during examination. When examining the drops always use the same light brightness, and do not use a polarizer because precipitate color is an important indicator (dark = bad sign, light/transparent = good sign). Examine and store the experiments at the same temperature to prevent irreproducible results.

13.2. Examination of the Crystallization Experiments

In order to save space in the lab book, and for easy identification, a score of one or more numbers is used to classify the contents of the drop. These scores (shown in parentheses) are presented in the legends to Plates 1-20.

Example: A score of 49 indicates gelatinous precipitate (score 4) with crystals (score 9). The scoring is not based on quality; a higher number does not necessarily mean a better result.

Check the entire depth of focus in the drop, first at 40 to 50× magnification and then more closely at 100×. Use polarized light to check the optical properties of crystalline material in the drop. Crystalline material, as opposed to amorphous precipitate, is birefringent, i.e., it splits (polarizes) transmitted light into two orthogonal beams. Birefringent precipitates, like crystals, sparkle or glisten.

Be sure to remove the tray from the observation platform before applying and removing the polarizer or analyzer. Without the tray, adjust the two polarizers so that the the field of illumination is uniformly dark. If a crystal is birefringent, some of the light passing through the crystal will be transmitted by the analyzer. The intensity of the transmitted light passes through maxima and minima when the crystal is rotated. This effect is not always clearly visible because of interference from the plastic tray with the polarized beam. A birefringent crystal in a plastic tray will have a different color than the background. (see Plate 20). Birefringence is a property of both protein and salt crystals and is even pronounced for salts.

- **The protein is not precipitated. (Plates 1 to 3)**

   Drop is clear. (Score=0) **Plate 1.**
   - Check the entire depth of focus in the drop; try dark-field illumination to check for gelatinous precipitate.
   - When screening for crystallization conditions, if the drop remains clear for more than two weeks, repeat the condition with a higher protein or precipitant concentration.
   - If limited amount of protein is available, place the clear drops over a reservoir with a higher precipitant concentration.
   - In an optimization step, clear drops can be seeded with crystals from neighboring drops.

   Drop contains non-protein particles. (Score=1) **Plates 2 and 3.**
   - Glass has an irregular shape without birefringence. **Plate 2.**
   - Dust on the cover slips, in and around the drop: Clean the cover slips carefully with a tissue before setting up the drops.

- Dust in the drop, but also in the well: Filter the stock solutions.
- Fibers from clothes. Fibers are colorful under polarized light. **Plate 3.**
- Bacterial growth: Add sodium azide to the protein and well solution.

- **Precipitate shows no birefringence and has no edges.** (Score=2-5) **Plates 4-10**

Drop is mostly clear but contains precipitated protein. (Score=2) **Plate 4.**

- This could indicate that the protein or precipitant concentrations (or both) do not favor nucleation and crystal growth.

The protein is fully precipitated with a dark color. (Score=3) **Plates 5-7.**

- Identify unfavorable conditions (pH, precipitant, or additive) by comparing the results from the sparse matrix or grid screens. These conditions should be avoided in future experiments as should any pH where the protein denatures regardless of precipitant type or concentration.
- If the protein precipitated within 1 day, repeat the condition with half the precipitant concentration. With viscous solutions like PEG the protein can precipitate when the protein and well solution are mixed. The drop will contain regions of heavy precipitate and regions that are clear. **Plate 5.**
- If the heavy precipitate is formed after a few days, the condition does not favor crystallization. **Plate 6.**
- In older drops a skin of denatured protein can appear on the surface of the drop. **Plate 7.** If the surface has a film that is not wrinkled it could be phase separation.

Gelatinous protein precipitate. (Score=4) **Plates 8 and 9.**

- Gelatinous precipitate is sometimes hard to see because it is white or transparent. **Plate 8.** Check the drop with dark-field illumination: the structural elements appear bright on a dark background.

- Gelatinous precipitate is a good starting point, when no crystals are found. Optimize the condition by changing one of the parameters (pH, precipitant concentration, or additive). **Plate 9.**

Phase Separation. (Score=5) **Plate 10.**

- The protein is visible as droplets within the drop or as an "oil" film on the surface of the drop. **Plate 10.** This can be seen best at the interface of the drop and cover slip. Temperature has an important influence on the equilibrium, so be careful not to heat the drop during examination. Sometimes crystals grow in the highly concentrated protein phase of the phase separation. Mounting crystals from this protein phase can be a problem, because of the unknown composition of the protein phase surrounding the crystal. Try cryo-freezing the crystal directly from the drop.

- **Precipitate shows birefringence or has edges.** (Score=6-9) **Plates 11-20**

Spherulites: transparent birefringent clusters. (Score=6) **Plates 11-14.**

- Spherulites can look like droplets, **Plate 11,** or like transparent clusters, **Plate 12.** When using polarized light, spherulites show dark and light parts. This is a good starting point for optimization.

- When crystals grow too fast, the edges disappear. **Plate 13.** Lower the protein or precipitant concentrations (or both) or increase the volume of the drop.

- Try streak seeding to optimize the conditions. **Plate 14.** Add an additive or inhibitor when the optimized condition doesn't yield crystals.

Small structures, where you might see edges. (Score=6) **Plate 15.**

- Microcrystals are small particles where edges and birefringence are visible if the magnification is high enough. **Plate 15.**

Crystal grown in one dimension. (Score=7) **Plates 16-18.**

- Needles, too small for data collection. **Plate 16.**
- Optimize pH and precipitant concentration.
- Try a lower protein concentration.
- Shower of needles grown from one nucleation point; try seeding. **Plate 17.**
- If the crystal is damaged during macroseeding, a lot of small crystals are formed. **Plate 18.**

Crystal grown in two dimensions. (Score=8) **Plate 19.**

- Try optimizing the conditions; use additives or seeding.
- When mounting plates for X-ray diffraction studies use capillaries with a large diameter, to avoid crystal breakage.

Crystals grown in three dimensions. (Score=9) **Plate 20.**

- Once the crystals are bigger than 0.05 mm, write down the dimensions (e.g., $0.4 \times 0.2 \times 0.1$ mm)
- Check for salt crystals in the well. If the drop also contains dark precipitate (score=3) then check if the conditions being used could possibly form salt crystals. Avoid phosphate.
- Mount the crystal and check diffraction quality.
- If the crystals are not suitable for X-ray diffraction analysis, try adding additives or inhibitors. Also go back to the initial screen and check for conditions that may lead to another crystal form.

14

Seeding

Enrico A. Stura

Plates 1-20. Crystallization drop phenomena and interpretation. Score in parentheses.

Plate 1. (0) Drop is clear.
Plate 2. (1) Piece of glass in the drop. Glass has an irregular shape.
Plate 3. (1) Fiber from clothes. Fibers are colorful in polarized light.
Plate 4. (2) Drop is mostly clear.
Plate 5. (3) Protein precipitated upon mixing with a viscous well solution. Visible as a clear region with dark edge.
Plate 6. (3) The protein is fully precipitated. The color of the precipitant is dark.
Plate 7. (3) Drop with wrinkled skin of denatured protein.
Plate 8. (4) The drop contains gelatinous protein precipitate.
Plate 9. (4) Gelatinous protein (polarizer used).
Plate 10. (5) Phase separation, visible as small droplets.
Plate 11. (6) Spherulites.
Plate 12. (6) Transparent cluster.

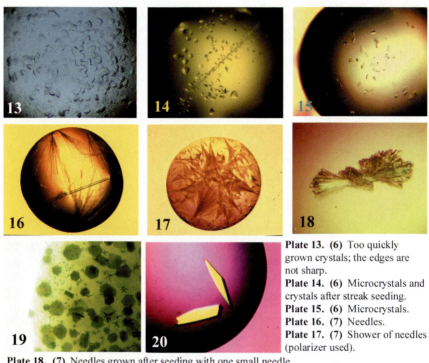

Plate 13. (6) Too quickly grown crystals; the edges are not sharp.
Plate 14. (6) Microcrystals and crystals after streak seeding.
Plate 15. (6) Microcrystals.
Plate 16. (7) Needles.
Plate 17. (7) Shower of needles (polarizer used).
Plate 18. (7) Needles grown after seeding with one small needle.
Plate 19. (8) Hexagonal plates.
Plate 20. (9) Crystals (polarizer used).

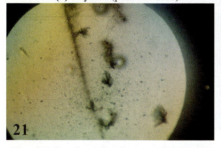

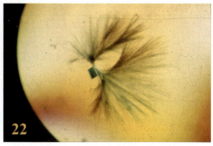

Plate 21. Streak seeding from one Fab steroid complex to a different one. The development of the streak line indicated that crystals of the first complex served as seeds for the second complex. Elsewhere in the drop sheaths of plate crystals nucleated spontaneously. Small prismatic steroid crystals are also visible. One of the sheaths appears along the streak line. This could have occurred by epitaxy. The poor quality of the crystals and the presence of two crystal forms complemented the information obtained from the IEF gel which indicated that the protein need further purification. After additional purification on a MonoQ (Pharmacia) column, crystals of X-ray quality were obtained under essentially identical crystallization condition.

Plate 22. Epitaxial nucleation. Stoichiometry is an important variable in the crystallization of complexes. Three different crystal morphologies were obtained as the stoichiometry was varied, starting with an excess of the first component and moving to an excess of the second. In this example, the second crystal form that nucleated epitaxially on the surface of the first crystal is the result of seeding between drops set up at different stoichiometries.

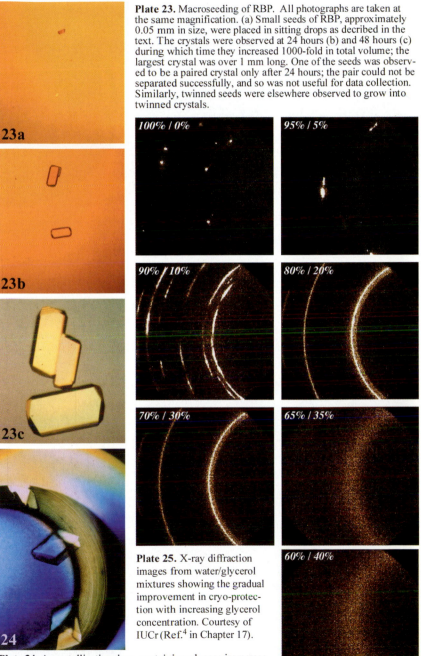

Plate 23. Macroseeding of RBP. All photographs are taken at the same magnification. (a) Small seeds of RBP, approximately 0.05 mm in size, were placed in sitting drops as decribed in the text. The crystals were observed at 24 hours (b) and 48 hours (c) during which time they increased 1000-fold in total volume; the largest crystal was over 1 mm long. One of the seeds was observed to be a paired crystal only after 24 hours; the pair could not be separated successfully, and so was not useful for data collection. Similarly, twinned seeds were elsewhere observed to grow into twinned crystals.

23a

23b

23c

24

100% / 0%

95% / 5%

90% / 10%

80% / 20%

70% / 30%

65% / 35%

60% / 40%

Plate 25. X-ray diffraction images from water/glycerol mixtures showing the gradual improvement in cryo-protection with increasing glycerol concentration. Courtesy of IUCr (Ref.[4] in Chapter 17).

Plate 24. A crystallization drop, containing glucose isomerase crystals, under oil. The outer rim is the boundary of the drop; the inner rim is the bottom of the crystallization plate.

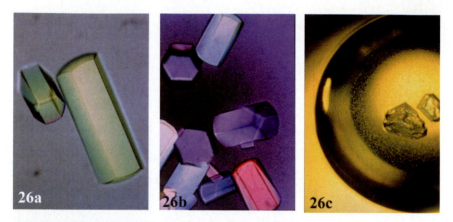

Plate 26. Appearances can be deceiving. (a) and (b) show PDGF (platelet-derived growth factor) crystals that diffract only to approximately 6 Å. (c) shows crystals of glyoxalase I that diffract to 1.7 Å.

Plate 27. Crystals of CRABP I (cellular retinoic acid-binding protein) complexed with its native ligand, all-trans retinoic acid.

Plate 28. Crystals growing from a phase separation.

14

Seeding

Enrico A. Stura

DIEP, CEA Saclay, France

Q. How do you get an elephant up a tree?
A. Get the elephant to sit on a seed and wait for the tree to grow.

Seeding is not a joke when it comes to macromolecular crystals. A seed provides a template on which further molecules can assemble, and given the proper environment, time, and patience, the seed will enlarge into a crystal.

The size of the final crystal is not dependent on the size of the initial seed but it will inherit many of the other characteristics of the seed from which it originated. Seed-grown crystals avoid some of the problems inherent to spontaneously nucleated crystals because spontaneous nucleation requires a higher degree of supersaturation. The higher supersaturation promotes other aggregation events, reversible and irreversible, which compete for protein and lower the degree of supersaturation. When supersaturation is reduced, the chance of forming a stable nucleus is lowered, and so crystals may never appear even under excellent growth conditions.

Spontaneous nucleation is statistical in nature: Its probability increases with the degree of supersaturation. In reverse screening, (see Chapter 11, **rule 1**, and reference[1]), we select a part of the solubility curve where there is a high degree of supersaturation to increase the probability of nucleation. Some crystal forms will not be easily accessible by this approach since their nucleation only occurs at high precipitant and low protein concentrations, and hence also at a low degree of supersaturation. This part of crystallization space is difficult to access without waiting long periods of time for spontaneous nucleation to occur. It is profitable to stimulate nucleation by seeding with crystals of the same protein grown under other conditions. Energetically, it is more favorable to add to an already existing crystal plane (even epitaxially) than it is to create a new nucleus, so seeding increases the probability of nucleation of a new crystal form.

This seeding method is called epitaxial jump[2] and it is used to obtain crystals under conditions which may be significantly different from those in which the original seed was obtained. Streak seeding, seed transfer, and variations of environmental conditions are utilized for epitaxial jumps.

Two pieces of equipment are needed for the seeding methods described here: a streak seeding[3] probe or wand, e.g., a cat whisker, and a crystal transfer syringe (Figure 14.1, a and c). One can learn to use these tools and practice the methods by repeating Experiment 11.1 in Chapter 11.

14.1. Streak Seeding

The whisker is used to gently touch a crystal and dislodge seeds (Figure 14.1, b, step 2). The seeds remain attached to the whisker and are transferred to a new protein-precipitant drop as the whisker is drawn in a straight line across it. Enough seeds remain on the whisker to seed 3-6 drops. Since successive drops receive less seeds, this is a method to titrate the seeds. Although little precipitant accumulates on the whisker, the precipitant dries out as the whisker travels from one drop to the next. This produces a local increase in precipitant concentration along the streak line, and it can stimulate nucleation in the absence of seeds. It is important to re-seal both the source and the receiving wells promptly after the transfer.

While crystals nucleated from seeds grow along the streak line, self-nucleated ones grow at other places in the drop (see Plate 21). Lack of a streak line indicates:

1. a too brief pre-equilibration time,

2. too low a precipitant concentration, or

3. incompatible crystallization conditions.

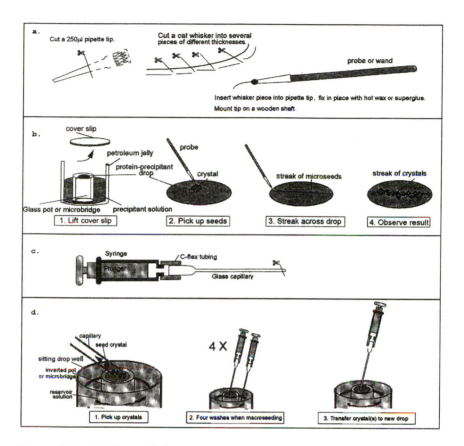

Figure 14.1. Seeding techniques:

a. Making a probe for streak seeding. With scissors or a razor blade, cut a pipette tip, insert a whisker into the end, and seal with wax or superglue. Trim the whisker to size; use the rest of it to make probes of different thickness. A 5 cm wooden shaft (e.g., a bamboo skewer) can be used to make a handle.

b. Streak seeding method. Open a well and gently touch a crystal with the whisker to pick up seeds. Streak the whisker across a pre-equilibrated, though not necessarily fully equilibrated, drop. Depending on the degree of supersaturation, the streak may become visible in a few seconds or after several days. The shorter the time the well remains open, the better the result will be. Hanging drops are prone to drying out and are not recommended.

c. Making a seed transfer syringe. Smear petroleum jelly on the shaft of a glass syringe. Place a piece of C-flex tubing (internal diameter 3/32") on the syringe tip. This will hold a standard X-ray glass capillary.

d. Transfer of seeds. Open a well (1) and pick up one or more seed crystals with a minimal amount of mother liquor. Transfer to a protein-precipitant drop (3). When the conditions have been optimized, macroseeding with four intervening washes (2) can be used. Fill four sitting drop holders with reservoir solution and place distilled water in the outer well. Transfer a single crystal through the four washes and rinse the capillary in between each wash with the distilled water.

The first two possibilities can be eliminated without setting up new drops: in (1) by repeating the seeding after further equilibration, and in (2) by successively adding extra precipitant to the reservoir, allowing time for the drop to re-equilibrate and seeding again (Table 14.1).

14.2. Preparing Seed Stock

Seed stock is made by crushing three or four crystals in a 1 ml glass tissue homogenizer with the addition of a stabilizing solution (mother liquor) to prevent the seeds from dissolving. Further solution is added to wash the crushed seeds from the sides of the homogenizer. The solution is transferred from the homogenizer to a test tube for storage of the seed stock. About 1 μl of the seed stock is diluted in 1 ml stabilizing solution and mixed. Further 10^{-1} to 10^{-4} dilutions are made and tested experimentally. The temperature must be kept constant as seeds may dissolve either on heating or cooling. One of the diluted seed solutions should be suitable to stimulate the growth of a small number of crystals in each drop. The seed solutions can be stored for up to 1 year. Add 0.1% azide to the stabilizing solution to prevent microbial growth.

The seed stock and seed dilutions can be used for streak seeding by dipping the whisker into them and streaking new drops (Figure 14.1, b, step 3). Start with the most dilute solution first as seeds are carried on the whisker for several rounds, and finish with the seed stock.

Experiment 14.1: Streak seeding and seed transfer

Purpose: (1) To demonstrate that seeding can eliminate the need for spontaneous nucleation. (2) To practice the basic techniques of streak seeding and seed transfer.

Use siliconized cover slips under the microscope.
   (i)   Follow steps 1, 2, and 3 in Chapter 11, Experiment 11.1, but use solutions stored at 4°C as these will give you large crystals more

rapidly than room temperature solutions. As soon as you can identify a crystal, set up an identical drop by repeating step 3. Place a drop of lysozyme solution on another cover slip and add an equal drop of the precipitant under a microscope. With the crystal wand touch a visible crystal from the first drop. Stir the new drop with the crystal wand. Make a note of the time it takes for the first crystals to appear and note the number of crystals produced.

(ii) Repeat the experiment, but this time do not stir the drop. Use the seeding syringe (Figure 14.1, d) to get some crystals from the first drop setup. Add the crystals (seeds) to the unstirred drop.

In step (i) we used a microseeding technique, and so we were unable to see the seeds attached to the whisker. In step (ii) we could see the seeds enter the capillary. This is called seed transfer, a macroseeding technique. The term macroseeding also refers to the same procedure with interposed seed washing steps (Figure 14.1, d, step 2).

(iii) Repeat the experiments in steps 4, 5, and 6 in Chapter 11, Experiment 11.1, and apply either streak seeding or seed transfer.

14.3. Diagnostic Uses of Seeding

Seeding provides information about the degree of supersaturation in the protein-precipitant drop. By seeding at different times during the course of equilibration, it can be determined when supersaturation is achieved. A seed transferred into an undersaturated solution will dissolve, confirming that the drop is undersaturated. Knowing the saturation status of the drop is a guide to optimization of the parameters in question (such as precipitant/protein concentration, etc.).

Seed transfer can be used to find the suitable preliminary seeding conditions before introducing the four washes of a good seed crystal. The washes are tedious but essential as they eliminate possible transfer of microseeds from the mother drop. However, they can also introduce other errors. The increased manipulation to which the seed crystal is subjected may damage it, also causing microseeds. The precipitant concentration of the wash may be too low or too high. Therefore, it is expedient to pre-test the conditions by seed transfer before investing the effort involved in washing the intended seed crystal.

14.4. Seeding and Sitting Drops

A sitting drop setup provides a better environment for seeding than a hanging drop since the well needs to be opened when seeding. Because of the drop's position at the center of the well, a sitting drop suffers less disruption than does a hanging drop, which is exposed to the relative dryness of the outside environment.

Temperature is an important factor which affects equilibration and nucleation rates and must be kept constant in order for seeding to be reliable. Temperature is better controlled by the use of a sitting drop vapor with glass pedestals[4] and a constant temperature room or incubator. Glass pedestals are preferable to plastic microbridges because: 1) the thermal conductivity of glass ensures that the protein drop is at the same temperature as the reservoir, and 2) glass is not as easily scratched.

The sitting drop method also avoids the problem of condensation on the glass cover slip caused by temperature changes. In a hanging drop experiment, the drop is effectively in thermal contact with the outside air and not with the reservoir, hence distillation can occur. This is very troublesome in low salt and PEG crystallizations where one may observe an increase rather than a decrease in the volume of the protein-precipitant drop as equilibration progresses. The sitting drop method reduces such problems.

14.5. Optimization

During screening, it is better to maintain a high degree of supersaturation to obtain results more rapidly. The degree of supersaturation will be decreased later to optimize the conditions for macroseeding. Before macroseeding, increase the drop size and slightly decrease the protein or precipitant concentrations or both. The larger drops mean supersaturation will be reached more slowly, and seed transfer or streak seeding can be used to determine when it is best to macroseed. Often good macroseeding conditions are established at reduced precipitant concentration but at a compensating higher protein concentration.

When optimizing conditions for growing crystals, remember that the best seeds come from the best crystals and result in the best crystals. Seeds from crystal corners may provide faster growth than seeds from crystal faces.

14.6. Seeding as a Screening Technique

Changing crystallization conditions Seeding avoids the need for spontaneous nucleation and therefore is well suited for testing minor and major changes to growth conditions, such as adding a co-precipitant, testing new additives, or scanning the pH range. Conditions that might be bad for spontaneous nucleation can be excellent for crystal growth. The seeded crystal will not have to compete with unseeded "weed-crystals." Sometimes even small changes result in important improvements in crystal quality.

Crystallinity assay Can you distinguish a microcrystalline precipitate from an amorphous one? Streak seeding can. A microcrystalline precipitate is a source of seeds. Often such precipitates are the results of too much precipitant or sub-optimal pH. Reduce the precipitant concentration, change the pH, and streak seed.

Crystallization of complexes When crystallizing complexes (between two proteins, protein and ligand, or protein and nucleic acid, etc.) there will be heterogeneity in the system. Apart from the heterogeneity of each protein, the resulting solution will be a mixture of complexed and uncomplexed molecules. In some cases it may be possible to pre-purify the complex, in other cases the stoichiometry must be varied to compensate for errors in protein concentration measurements. Here are some general guidelines:

1. For protein complexes with small ligands adding an excess of ligand, typically 1:1 to 1:20, is sufficient. Larger excesses of ligand may even inhibit crystal growth.

2. In the case of enzymes, inhibitors are preferable to substrates, as catalysis of the substrate into product will increase the heterogeneity in the system.

3. Tri-phosphated nucleotides, such as ATP, are easily hydrolyzed. Hence it may be advantageous to use non-hydrolyzable analogues such as AMP-PNP, adenosine diphosphate-γ-S (ADPγS), and their analogous guanosine derivatives instead.

Streak seeding can be used to optimize the ratio and concentrations of the proteins from the response along the streak line. Different crystal forms may be promoted by the use of different stoichiometries (see Plate 22). The initial microcrystals or microcrystalline aggregates can be used to start the procedure.

14.7. Heterogeneous Seeding

Surfaces promote adhesion because it is energetically easier to add to a surface than to create a new nucleus or site of aggregation. In crystallization, this is more often a problem than an advantage. Glass surfaces used to grow crystals are siliconized but still crystals are found attached to them. The interaction with the glass can be stronger than the internal lattice forces so that the crystals break when an attempt is made to dislodge them.

On the positive side, adhesion to a regular surface may provide a suitable template to start an ordered protein layer, leading to the nucleation of a crystal. Epitaxial nucleation occurs at a lower degree of supersaturation than spontaneous nucleation. For example, mineral substrates have been used for epitaxial nucleation of protein crystals,[5] and epitaxial nucleation can also occur on cellulose fibers present (often unintentionally) in the protein-precipitant drop. Hence, streak seeding with promising aggregates or precipitates may succeed not necessarily because the aggregates are crystalline, but because they provide an ordered plane on which to nucleate.

Experiment 14.2: Epitaxial jumps

In Experiment 14.2 below we will explore ways in which we can change from one crystal form to another.

Seeds can be used to induce nucleation in supersaturated solutions equilibrated under conditions quite dissimilar from those in which the seed crystals originated. The new crystals may have a lattice related, but not identical, to that of the seeds.

A. Purpose: To stimulate the growth of a different crystal form by changing the precipitant concentration after crystals have grown.

(i) Set up drops with equal volumes of lysozyme solution (100 mg/ml in 50 mM sodium acetate, pH 4.5) and 30% MPEG 5000 in 100 mM imidazole malate. Let mixing occur by itself. Add a small crystal (Figure 14.1, d) grown in 30% MPEG, 1 M NaCl, 50 mM sodium acetate pH 4.5 (as obtained in Chapter 11, Experiment 11.1). Some growth should be apparent within fifteen minutes. Allow this crystal to grow overnight. New crystals will also nucleate and grow during this time. The next day most of the growth will have occurred and the level of supersaturation in the drop should be substantially lower.

(ii) Increase the concentration of the precipitant in the reservoir. This can be done either in a series of gradual additions or one large jump in concentration. After each addition of precipitant, observe the drops for the appearance of new crystals. This may take from a day to a week. These are likely to nucleate off the original crystals and grow in a different morphology and direction from them. An epitaxial jump may have occurred. Use streak seeding in new drops to propagate the new crystals.

B. Purpose: To stimulate the growth of a different crystal form by streak seeding into different crystallization conditions or changing the crystallization conditions after seeding as in A.

(i) Set up drops under any desired supersaturated conditions. Different protein and precipitant concentrations should be tried. To obtain more tightly packed crystals, the protein concentration should be halved and new supersaturated conditions established at higher precipitant concentrations.

(ii) Streak seed drops after equilibration with any crystal form obtained for the same macromolecule. The seeds themselves may not be able to grow in the new conditions, which are typically characterized by higher precipitant and lower protein concentrations, but they may serve as templates that enable a switch to a more compact crystal form.

(iii) Check for the development of a streak line. If a line develops, use the crystals obtained to seed other drops. If growth is slow, increase the precipitant concentration.

(iv) Propagation of epitaxially grown crystals:

Use seeds from either **A** or **B** to seed new drops. Once crystals can be grown large enough, check if their diffraction limit is improved. Crystals which have been grown in this manner and have a lower solvent content are termed **squeezed** crystals.

14.8. Recognizing an Epitaxial Jump

Sometimes crystals that result from a jump may be similar in shape to those from which they derived. For example, a new crystal form of the tissue factor-factor VIIa-5L15 complex (Tf.VIIa.5L15)[6] resulted from streak seeding followed by an increase in precipitant concentration (Figure 14.2). The new crystal form had a different habit but was morphologically very similar to the original crystals.

14.9. Cross Seeding

Cross seeding between wild type and mutated protein, or between uncomplexed and liganded or chemically modified proteins, can be tried in the same manner by epitaxial nucleation. Because solubility is often different for the "changed/mutated" protein, it may not crystallize

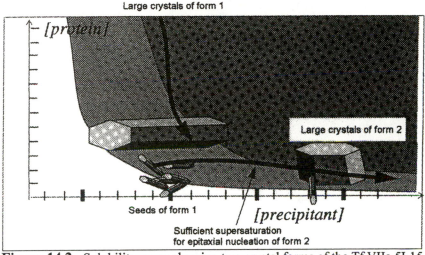

Figure 14.2. Solubility curve showing two crystal forms of the Tf.VIIa.5L15 complex. Large crystals of form 1 (maximum resolution 7Å) were grown at high protein, low precipitant concentrations. To grow crystals of form 2, a new drop with a lower protein and higher precipitant concentration was equilibrated and seeded with the form 1 crystals. The drop became sufficiently supersaturated for epitaxial nucleation of form 2 crystals to occur on the face of the form 1 type. The new crystals were somewhat thicker and larger but otherwise similar in appearance. Analysis by X-ray diffraction showed that the new crystals had the same space group, but different cell axes and a tighter packing. The diffraction improved to 3.2Å. See reference[6] for more details.

in the same conditions as the native protein. However, it may do so when a much wider range of conditions are tried in conjunction with seeding. Seed transfer and streak seeding from the native protein crystal will provide both a multi-plane environment for epitaxial nucleation and larger crystals than microseeding. If the transferred crystals dissolve, the precipitant concentration may be raised to compensate for solubility differences. If this does not help, then the crystallization conditions and seeds are incompatible.

14.10. Summary

Seeding methods provide faster means to obtain results than conventional methods. Many more conditions can be analyzed with the same amount of protein. Streak seeding allows for easy visualization of the results whether used for homogeneous or heterogeneous seeding. The application of micro- and macroseeding methods make growing large single crystals for X-ray structure determination less of a haphazard endeavor, and many projects would not have been viable without it.

Q. How do you get an elephant up a tree?
A. Teach it to jump! (epitaxially).

References and Further Reading

1. Stura EA, Satterthwait AC, Calvo JC, Kaslow DC, Wilson IA: **Reverse screening.** *Acta Cryst* 1994, **D50**:448-455.
    Describes the rules of reverse screening.

2. Stura EA, Charbonnier JB, Taussig MJ: **Epitaxial jumps.** *Proceedings of the 7th International Conference on the Crystallization of Biological Macromolecules*, 1998, 208.
    Describes several cases of epitaxial jumps.

3. Stura EA, Wilson IA. 1991. **Application of the streak seeding technique in protein crystallization.** *J Cryst Growth* 1991, **110**:270-282.
    Gives several examples in which seeding is used to enhance crystal quality.

4. Stura EA, Johnson DL, Inglese J, Smith JM, Benkovic SJ, Wilson IA: **Preliminary crystallographic investigations of glycinamide ribonucleotide phosphorylase.** *J Biol Chem* 1989, **264**:9703-9706.
    Introduction of the sitting drop method with inverted glass pots.

5. McPherson A: **Macromolecular crystals.** *Scientific American* 1989, **260(3)**:62-69.
    Epitaxial nucleation on a mineral substrate. Easy reading about protein crystallization.

6. Stura EA, Ruf W, Wilson IA: **Crystallization and preliminary crystallographic data for a ternary complex between tissue factor, factor VIIa and a BPTI-derived inhibitor.** *J Cryst Growth* 1996, **168**:260-269.
    First published case of an epitaxial jump. Crystallization screening with 1.5 mg of protein. Calcium and NaCl were used to modulate protein solubility.

A 14.1

Table 14.1. Recommendations for streak seeding and seed transfer

Nucleation Method	No crystals/No line	Crystals along line	Crystals everywhere
Spontaneous	Streak seed with crystals grown under other conditions. Precipitate present in the drop will be cleared by the seeds. If no precipitate is present, streak seed and increase precipitant concentration by adding salt or extra precipitant to reservoir.	Drop may have been streak seeded unintentionally: crystals do not grow spontaneously along a straight line.	Use crystals for seeds irrespective of their quality. Set up more experiments at lower precipitant concentration; if they do not nucleate spontaneously, streak seed the drop.
Streak Seeding	Increase precipitant concentration or add salt to reservoir, allow to re-equilibrate and then re-seed.	Positive result. Explore different precipitants, buffers, pH, additives, temperatures to obtain better crystals.	Supersaturation of drop is too high, set up drops at lower precipitant concentration.
Seed Transfer	Drop is undersaturated. Increase precipitant concentration or add salt to reservoir and re-seed.	Capillary may have been moved in a straight line while transferring crystals. Introduce wash cycles (Figure 14.1, d).	Microseeds may be responsible. Introduce 4 cycles of washing (Figure 14.1, d)

15

Macroseeding: A Real-Life Success Story

Sherry L. Mowbray

15

Macroseeding: A Real-Life Success Story

Sherry L. Mowbray

Swedish University of Agricultural Sciences,
Uppsala, Sweden

15.1. Why Do Macroseeding?

We never do macroseeding willingly, but have been driven to it a number of times. The usual situation is one where protein is reasonably plentiful, but extensive trials with other methods have resulted only in small or poor crystals. As in other types of seeding, the object of macroseeding is to disconnect the nucleation phase from that of crystal growth, that is, to find conditions where growth will occur but nucleation will not. By introducing a seed crystal under these conditions, we try to force it to grow in preference to the formation of new crystals. The most fundamental difference between macroseeding and other types is that, in macroseeding, you can see the seed—choosing a good one is a critical factor in final crystal quality. The results can be well worth the relatively large amount of manipulation required.

To be successful with macroseeding, it is obviously necessary to characterize the conditions that favor the things you want and disfavor the things you don't. It is most efficient to build a macroseeding procedure in several discrete steps, each of which utilizes the results of the previous ones. The rationale and typical protocol for each step are outlined below with *Escherichia coli* ribose-binding protein (RBP) as an example. (The macroseeding of RBP was originally worked out by Brent Cole while we were at the Howard Hughes Medical Institute in

Dallas, Texas, USA, and has since been refined by Joakim Björkman at the Swedish University of Agricultural Sciences, Uppsala, Sweden.)

15.2. How to Macroseed

Step 1. Obtain seeds. It is extremely important to use single seed crystals; absolute size is not nearly as important as absolute quality. The series of pictures shown in Plate 23 (a, b, and c) for RBP illustrate both the dramatic results obtained when things go right as well as an important lesson about how things can go wrong. I have seen damaged crystals repair broken edges during growth, but never a twinned crystal become single—macroseeding will usually only worsen any twinning or other irregular features of the seed.

General protocol: Prepare enough (30 or more) usually small (50-100 μm) seed crystals by whatever method is convenient. For example, repeated streak seeding can be a good way of getting higher quality seeds, and who knows, you might not have to go further than that.

Example: For RBP, poor quality crystals could be obtained by batch crystallization using 3.5 to 7 mg/ml RBP, 10 mM sodium phosphate, pH 7.0, 10 mM ribose, 25% PEG 4000 (Merck), or by vapor diffusion under similar conditions. Small single seeds could be broken off bundles of crystals, or occasionally, found. The seeds could sometimes be improved by microseeding of additional drops that were then equilibrated by vapor diffusion.

Step 2. Determine the solubility of the crystals. The goal of this step is to find the lowest concentration of precipitant at which your crystals are stable (without added protein), to use for washing as well as for designing the actual crystallization trials.

General protocol: Test a series of precipitant solutions, ranging from about 50% of the precipitant concentration needed for crystals to appear, to about 150% of same. Don't forget to keep appropriate buffering, additives, etc., as these will often affect crystal stability. Crystals (which need not be of high quality) will be added to these different solutions, and observed to find conditions where: a) they are stable for about 15 minutes, but not more than 1 hour, and b) they are stable indefinitely.

Example: Solutions of 30%, 25%, 20%, 19%, 18%, 17%, 16%, and 15% PEG 4000 are mixed in 10 mM sodium phosphate, pH 7.0, and 10 mM ribose. One ml of these solutions is placed in separate wells of a 9-well spot plate. A small crystal of RBP is added to each drop, which is then sealed individually with a cover slip and silicone grease. Observe at intervals; don't leave the plate on the microscope platform in between observations, as the lamp will warm up the plate, even after it is turned off. RBP crystals last for days at 20% PEG 4000, a few hours at 18%, and 30 minutes at 16%, but melt instantly at 15%.

Step 3. Generate a wash protocol. The purpose of washing the crystal is both to remove extraneous nuclei and to melt the seed crystal slightly (thus providing a fresh surface for crystal growth).

General protocol: First do a trial wash series, transferring a single crystal from high to low concentration of precipitant (without protein). Start with a concentration where crystals should be stable indefinitely, and end with a concentration that would melt them if they were left there. If necessary, modify the concentrations used on either end of the concentration range, to keep the crystal intact. Large wash drops work more efficiently, while smaller drops make it easier to keep track of the crystals; the volume you use in each drop (50-1000 μl) will depend largely on the containers you have and the size of your seeds. If your drops are small, you will probably need to increase the number of washing steps to get the same final effect.

Example: In a spot plate arrangement similar to that described above, transfer an RBP seed through the same series of decreasing PEG concentrations. The crystal is placed at the top of each drop, allowing it to fall through the solution to the bottom; this improves the washing process by physically dislodging small nuclei. The entire sequence takes 30 minutes or so. RBP seeds will last for 5 minutes in the last well, but not longer. So, during the actual seeding (next step), one can use the 18-20% drop as a holding tank, moving the seeds down to the 15% drop just before they are needed. Multiple seeds can easily be prepared at the same time, in the same spot plate.

Step 4. Establish appropriate equilibration conditions. By methodically testing different combinations of initial and final conditions in vapor diffusion experiments, it is often possible to see patterns that lead to effective crystal growth without new nucleation.

General protocol: Set up a grid of different initial and final precipitant concentrations, usually as sitting drops. (Although hanging drops can be used, most procedures are easier with sitting drops.) Remember that a protein is a significant component of the mixture. Normally, added protein will stabilize crystals at any given precipitant concentration so the initial conditions (in the drop) should include precipitant concentrations even lower than those used in the wash. The final concentrations of precipitant (used in the reservoir) should cover a slightly higher range. An informative grid of possibilities would look something like this (where i is the lowest precipitant concentration used in the wash):

initial precipitant concentrations: i - 20%,   i - 10%,   i
final precipitant concentrations: i,   i + 10%,   i + 20%

Three different starting concentrations are tested versus three final concentrations to give a total of nine possible combinations. The protein concentration in each drop can often be as low as 2 mg/ml (to reduce nucleation). Use 20 µl drops, unless you can really afford more protein. You will need to prepare a washed seed for each experiment, using the procedure established above. When the seed is almost ready (in the holding tank), prepare the sitting drop; then finish washing the seed, and add it immediately to the drop. It is important that a minimum volume of wash solution should be introduced into the crystallization drop, to keep control of the conditions. Also, use a fresh capillary (or loop) or wash the one you are using in deionized water just before the last transfer (to reduce the accidental introduction of microseeds). Seal this experiment before proceeding to the next (to keep concentrations as you mixed them, rather than allowing evaporation to occur).

Score your results on day 1, day 2, and at other times up to 2 weeks. As soon as you know what worked and what didn't, you'll probably want to refine the conditions by adjusting the initial and final precipitant conditions, protein concentration, size of vapor diffusion vessel, etc.; see below for some ideas. When the final protocol is worked out, it usually works consistently.

Example: Prepare three 20 µl sitting drops, 13-15% PEG 4000 (in a 3-well spot plate), 5 mg/ml RBP in 10 mM sodium phosphate, pH 7.0, and 10 mM ribose; mix thoroughly. Place a washed seed in each drop. Put the spot plate in a small sandwich box containing a reservoir solution of 15-20% PEG 4000, seal with parafilm and allow the solutions to equilibrate. RBP crystals grow over 4 days to 1 week, and have to be harvested before they crack or grow too big to be mounted in X-ray capillaries.

Step 5. Analyze results: Macroseeding rarely works perfectly the first time; it is most often the thoughtful analysis of the initial results that makes the difference between success and failure.

15.3. Common Problems/Causes

15.3.1. Too Many Crystals
1. improperly washed seed (extra nuclei)
2. initial precipitant concentration too high in sitting drop
3. final precipitant concentration too high in reservoir
4. diffusion too fast (use a larger vapor diffusion vessel or lower the temperature to slow the rate of equilibration)

15.3.2. Dissolving Seeds
1. precipitant concentration too low in either drop or reservoir
2. concentration of precipitant too low in final wash
3. too much of final wash solution added to the sitting drop

15.3.3. Poor Crystals
1. poor seeds
2. inappropriate equilibration conditions (try slowing things down

by lowering the temperature, using smaller jumps in precipitant concentration, or decreasing the protein concentration)

An interesting point here is that the rates at which the crystals grow in different directions are often altered by changing the parameters. By judicious choice of conditions, it may arrive at a more suitable morphology for data collection than that seen in the original seeds.

References and Further Reading

1. Thaller C, Weaver LH, Eichele E, Wilson E, Karlsson R, Jansonius JN: **Repeated seeding technique for growing large single crystals of proteins.** *J Mol Biol* 1981, **147**:465-469.

    Provides examples of crystallization of different proteins and some useful theoretical considerations.

2. Mowbray SL, Petsko GA: **The X-ray structure of the periplasmic galactose binding protein from Salmonella typhimurium at 3.Å resolution.** *J Biol Chem* 1983, **258**:7991-7997.

    Provides a description of my first (successful) macroseeding attempts.

3. Mowbray SL, Cole, LB: **1.7 Å X-ray structure of the periplasmic ribose receptor from Escherichia coli.** *J Mol Biol* 1992, **225**:155-175.

    More information about the ribose-binding protein and its crystallization.

16

Oils for Crystals

Naomi E. Chayen

16

Oils for Crystals

Naomi E. Chayen

Imperial College of Science, Technology and Medicine, London, U.K.

The crystal growth of proteins is a complicated process which is dependent on many factors, as seen by the variety of topics covered in this book. This chapter presents a novel approach which employs the use of oil as a major element to aid protein crystallization and which has established a new concept in the field of protein crystal growth. Oils are used (a) in screening to find initial crystallization conditions and (b) for optimizing crystallization by control of nucleation and growth. A step-by-step guide to the utilization of oils is given below.

16.1. The Rationale for Crystallization under Oil

Crystallization breaks down naturally into two phases: screening, where numerous different conditions are applied in order to obtain crystals of any description, and optimization, where one tries to improve the size and diffraction quality of the crystals. Since many of the more interesting proteins are often available in limited supply, there is demand for techniques which use minimal amounts of material. Reduction in protein consumption creates two major problems: (a) inaccuracy of dispensing, and (b) evaporation and drying out of the samples.

The development of a crystallization technique called microbatch overcomes these two difficulties by dispensing and incubating the samples under oil.[1] Moreover, this oil-based method provides a reliable environment for controlling the nucleation and growth stages of crystallization.

16.1.1 The Microbatch Technique

Microbatch is essentially a batch method in which the molecule to be crystallized is mixed with the crystallizing agents at the start of the experiment. The concentration of the ingredients is such that supersaturation is achieved immediately upon mixing, thus the composition and the volume of a trial remain constant and crystals will only form if the precise conditions have been correctly chosen. This is in contrast to other crystallization methods in which the protein solution is undersaturated at the outset of the experiment and conditions are changing from the time of setup until equilibrium is reached.[2,3] The objective of the microbatch technique is to reduce the consumption of sample by generating crystallization trials in volumes of 1-2 µl drops.

16.1.2 Setting up Crystallization Trials by Microbatch

Because it is difficult to accurately compose such small drops from many different components manually, an automated (computer controlled) micro-dispenser, called IMPAX, was designed (see references[1,2] and references therein). Obviously, maximum benefit is gained when experiments are set up automatically, but microbatch trials can also be performed manually.

Figure 16.1, a, illustrates the act of dispensing a crystallization trial under oil. A crystallization drop (which contains the protein to be crystallized and the crystallizing agents) is dispensed either automatically with IMPAX, or manually (using a standard micropipette) into a microtiter plate (Figure 16.1, b). To prevent evaporation of such small volumes, the crystallization samples are dispensed and incubated under the surface of a thick layer (>5 mm) of paraffin oil (for details, see Exercises 16.1 and 16.3). As the dispensing tip is withdrawn from the oil, the aqueous drop detaches from the tip and sinks to the bottom of the vessel since it is heavier than the oil. The tip is wiped clean by the oil, thereby preventing any carry-over from one trial to another if the same dispensing tip is used in the next drop. Mixing of the sample can take place either prior to dispensing or inside the oil, by stirring the drop with the dispensing tip.[3]

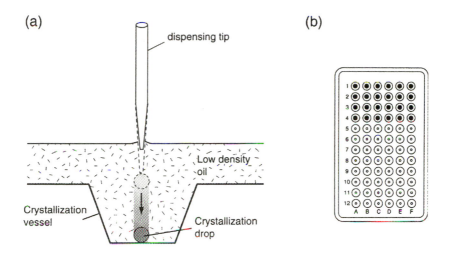

Figure 16.1. (a) The procedure of dispensing a crystallization trial under oil. The dashed circle represents the initial position of the crystallization drop at the time of dispensing. (b) Schematic diagram of a microtiter plate in which microbatch trials are incubated. The plate is 8×5.5 cm. The tray is filled with oil and the drops are dispensed into the oil. Black dots represent crystallization drops in wells; open circles represent empty wells.

Using microbatch, crystals of diffraction size and quality can be grown in 1-1.5 μl drops (with a robot) or 2 μl if dispensed manually. Plate 24 shows a 1.3 μl crystallization drop under oil containing crystals of glucose isomerase. (Protein was supplied by Dr. Robin Jackson.) The crystal in the center measures $0.50 \times 0.38 \times 0.12$ mm, confirming that the small size of the drop is not a limiting factor to attaining large crystals.

Most of the proteins tested in our laboratory could be crystallized under oil. The oils described here do not interfere with the common precipitants such as salts, polyethylene glycol (PEG), Jeffamine, and 2-methyl-2,4-pentanediol (MPD). Moreover, samples containing detergents have also been crystallized under oil.[2]

16.1.3. Crystallization of Membrane Proteins in Oil

A question which is often raised concerns the suitability of an oil-based environment for the crystallization of membrane proteins. The oil-based microbatch method was tested on two membrane proteins: photosystem I and chlorophyll binding protein 43 (CP43) of the photosystem II (PSII) from spinach. Crystals were obtained and it is possible that the oil played a critical role by slowly absorbing the detergent from the aqueous drop, thereby forcing the protein out of solution.[2, 3]

16.1.4. Additional Benefits of Oil

16.1.4.1 Protection from Physical Shock

External disturbances such as vibration can cause excess nucleation that leads to smaller crystals or crystal imperfections. The viscous oil in the microbatch plate buoys and cushions the crystals, making trials less susceptible to vibration. Moreover, unmounted crystals can be easily transported.

16.1.4.2 Resistance to Dissolution

In vapor diffusion one often observes increases in drop volumes due to slight variation in conditions such as temperature or the absorbance of volatile agents. This can cause dissolution of crystals. In microbatch experiments, provided the crystals are incubated under a sufficiently thick layer of paraffin oil, the volume of the drops remains constant and no dissolution occurs unless the solubility of the protein is temperature dependent.[2,3]

16.1.5. Limitations of Crystallizing under Oil

Not every case is suitable for crystallization under oil. A limitation of this method is that it cannot be applied where volatile organic molecules (e.g., dioxane, phenol, thymol) are required as precipitants or additives in the crystallization medium. However, organic substances which are not volatile (e.g., glycerol, tert-butanol, propanol) can be saturated into the oil prior to setting up the experiments.[2,3]

16.1.6. Harvesting Microbatch Crystals

Harvesting crystals from oil is somewhat more difficult than harvesting from cover slips or sitting drops since the crystals need to be "fished out" of the oil. Harvest solution usually needs to be added to the crystals before their removal from the oil. This is done by adding 15 to 30 μl of harvesting/stabilizing solution to the drop containing the crystals. As in the case of vapor diffusion, the harvest solution for microbatch should contain a slightly higher (about 5%) concentration of precipitant than that in the drop. If the crystals stick to the plate they are gently loosened (inside the drop), with microtools or with a whisker. After waiting awhile (up to 30 minutes) to allow the crystals to equilibrate, a standard micropipette is used to withdraw the enlarged drop from the oil and transfer it to a depression well containing more harvest solution. From that stage onward the mounting is continued as it would be done in a vapor diffusion trial.[2]

16.1.7. The Effect of Different Oils on Microbatch

The oil which was initially used for microbatch trials was paraffin liquid light ($\rho = 0.84$ g cm^{-3}), a purified mixture of liquid-saturated hydrocarbons obtained from petroleum. Paraffin was chosen after testing a variety of oils, many of which were not suitable due to their interaction with the crystallization trials (e.g., causing phase separation or precipitation).[3] The primary role of the oil was solely to act as an inert sealant to prevent evaporation of the small volume trials. However, experimental evidence soon revealed that the oil itself could play an important part in the outcome of the crystallization experiment, as described in the next section.

16.2. Use of Different Oils in Screening

Water evaporates at different rates through different oils. Paraffin oil permits only a negligible amount of water to evaporate, while silicone oils (polymers of repeating dimethylsiloxane units) allow it to pass freely. It is thus possible to manipulate the rate of evaporation by mixing paraffin and silicone oils in different ratios.[4]

If the trials are covered with a mixture of paraffin and silicone oils or simply with a thinner (<2-3 ml in a Terasaki plate) layer of paraffin oil, some evaporation of the trial occurs, thereby concentrating the sample with time. Using the paraffin/silicone mixture, crystals tend to appear within a shorter time than with paraffin oil alone.[4] However, such trials can dry out within a month, so frequent monitoring of the trials is imperative.

My preference is to apply paraffin oil to cover the trials for screening experiments, although it can be much slower in generating results. The effect of time as an additional factor was revealed when crystallizing β-crustacyanin, which could only be crystallized by the microbatch method under paraffin oil and took 4 months. No crystals were produced under a mixture of paraffin and silicone. In spite of being covered by paraffin oil, apparently during the lengthy time of incubation some evaporation was taking place, as there is no absolute immiscibility of oil and water. It seems that the β-crustacyanin solution underwent a very gradual concentration until it reached a point suitable for nucleation and subsequent growth of crystals.[2,3,5]

In summary, if one is impatient and short of time, use a mixture of paraffin and silicone oils for screening, with the risk that the trials could dry out after a month. On the other hand, trials under paraffin oil may produce crystals even after a year.

16.3. Use of Oils in Optimization

As for screening, optimization of conditions in order to improve the quality of crystals can be performed either manually or automatically. Optimization by means of controlling the nucleation or growth rates is mostly performed manually using paraffin or silicone oils for microbatch trials, and combinations of oils for vapor diffusion experiments.

16.3.1. Control of Nucleation

A major and common problem in protein crystallization is that often, due to excess nucleation, thousands of tiny crystals are formed instead of the desired few, large crystals.

In order to control nucleation one must work with very clean solutions. In microbatch where the drops are maintained under oil, the samples are never exposed to air and are therefore protected from airborne contamination. This makes the microbatch an ideal environment for controlled heterogeneous nucleation experiments.[2,3]

It is possible to set up totally clean trials by filtering the crystallization trials (containing a mixture of the protein solution and the crystallizing agents) through filters which allow the removal of particles as small as 100 nm from small volumes (see Exercise 16.2). Provided the trial remains under oil, nucleants can be inserted in a controlled manner since the oil prevents any other contaminant from entering the drop. Filtration of a crystallization trial through a 300,000 molecular weight cutoff (MWCO) filter just prior to setting up the experiment can prevent nucleation (no crystal formation at all) under conditions which would produce a substantial number of crystals if the solution is not filtered. Experiments have been performed in which the degree of nucleation, and consequently the number and size of protein crystals, was determined at will by the addition of different quantities of a nucleant to filtered trials. Nucleants can be seeds or any other external substance (see Exercise 16.2). The cleanliness of such trials has produced highly reproducible results.[3] Moreover, finding conditions for seeding is easier and more reliable in microbatch.

16.3.2. Effect of Surface Contact on Nucleation

Solid surfaces such as the walls of the container of the crystallization trial can also act as nucleants and cause excess nucleation, leading to the production of large numbers of low-quality crystals.

A series of microbatch experiments shown in Figure 16.2 demonstrate how the application of oil can determine the contact area between the trial and its supporting vessel, thereby enabling the experimenter to monitor the nucleation and reduce or increase its level at will. The figure illustrates three situations. In Figure 16.2, a, the drop has been dispensed onto the floor of a vial and then covered by a layer of oil; the drop spreads out and flattens over the floor of the container. Figure 16.2, b, illustrates a drop dispensed into oil as performed by the normal microbatch procedure (shown in Figure 16.1, a); the drop forms a spherical shape, with just a small part of it touching the floor.

Figure 16.2, c, represents a situation of "containerless crystallization[6,7]" in which a crystallization drop is suspended between two oils of different densities: one of higher and the other of lower density than that of water and the common crystallizing agents. The two oils, high density fluorinated silicone fluid ($\rho = 1.27$ g/cm^3) and low density polydimethylsiloxane ($\rho = 0.92$ g/cm^3), are not miscible and the drop floats at the interface, not touching the container walls (see Exercise 16.3). The number of crystals produced by procedures (b) and (c) is significantly reduced (by as much as 10-fold) and their size is, on average, 3 times larger compared with those grown by procedure (a), where the drop has the largest contact area with its vessel. Procedure (c) produces only marginally larger crystals than procedure (b), indicating that the interface between the two oils also acts as a surface but with reduced nucleation properties compared with that of a solid material.[3]

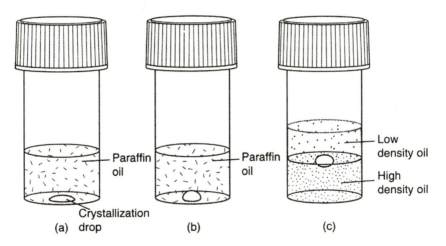

Figure 16.2. Application of oil for determining the contact area between a crystallization trial and its supporting vessel. (a) Large contact area. (b) Small contact area. (c) "Containerless crystallization."

16.4. The Use of Oil for Controlling the Rate of Vapor Diffusion Trials

The process of protein crystallization often takes place too rapidly, producing showers of small crystals instead of single large ones. A novel technique for slowing down the crystallization process without changing the crystallization conditions, or requiring any dedicated apparatus, was devised by inserting an oil barrier over the reservoir of conventional vapor diffusion trials (Figure 16.3).

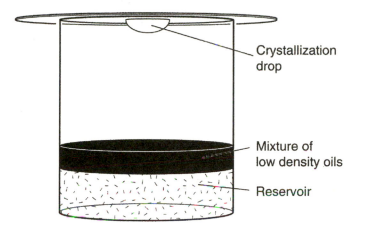

Crystallization drop

Mixture of low density oils

Reservoir

Figure 16.3. A hanging drop experiment with an oil barrier.

The oil barrier consists of a mixture of equal volumes of paraffin and silicone oils (see Exercise 16.4). It has been demonstrated that the thickness of the oil layer situated above the reservoir determines the rate of crystallization. Crystals of lysozyme and thaumatin grown in control trials containing no oil barrier appeared overnight with average crystal size measuring 0.1 × 0.07 × 0.05 mm. The crystals did not increase in size over a period of a month. In contrast, crystals grown under almost identical conditions, the only difference being that an oil layer of 2.5 mm was placed above the reservoir, appeared after 3 days and continued to grow over a period of 14 days to reach their full size of 0.5 × 0.3 × 0.2 mm. Using Linbro and Cryschem plates, the application of oil volumes between 200 to 700 μl (1-2.7 mm thickness) delays the crystallization time, thereby leading to the production of much larger crystals compared with control experiments which have no barrier and in which crystals grow over a shorter period of time.[8]

16.5. Summary

The use of oil has turned out to be an unexpected but a tremendously welcome tool in the crystallization of biological macromolecules.

In addition to the primary role of the oil in facilitating the crystallization of small volume trials, the oil can play a part in the control of nucleation, affect the rate of equilibration, and consequently determine the size of the forming crystals. Whether used for microbatch, vapor diffusion, or for control of nucleation, the presence of oil is a parameter which can contribute to the accuracy, cleanliness, and reproducibility of experiments. Furthermore, the oil has a protective role during the trials, and maintains the stability of the resulting crystals afterwards.[3]

References and Further Reading

1. Chayen NE, Shaw Stewart PD, Baldock P: **New developments of the IMPAX small-volume automated crystallization system.** *Acta Cryst* 1994, **D50**:456-458.

   Automation of microbatch and vapor diffusion.

2. Chayen NE: **Comparative studies of protein crystallization by vapor diffusion and microbatch.** *Acta Cryst* 1998, **D54**:8-15.

   When to use microbatch and when to use vapor diffusion.

3 Chayen NE: **The role of oil in macromolecular crystallization.** *Structure* 1997, **5**:1269-1274.

   Use of oil as a tool for crystallization.

4. D'Arcy A, Elmore C, Stihle M, Johnston JE: **A novel approach to crystallizing proteins under oil.** *J Cryst Growth* 1996, **168**:175-180.

   Screening by combination of microbatch and diffusion.

5. Normile D: **Search for better crystals in inner and outer space.** *Science* 1995, **270**:1921-1922.

   Growing crystals in oil and in microgravity.

6. Chayen NE: **A novel technique for containerless protein crystallization.** *Protein Engineering* 1996, **9**:No. 10, 927-929.

   Containerless crystal growth.

7. Lorber B, Giege R: **Containerless protein crystallization in floating drops: application to crystal growth monitoring under reduced nucleation conditions.** *J Cryst Growth* 1996, **168**:204-215.

Containerless crystal growth.

8. Chayen NE: **A novel technique to control the rate of vapor diffusion, giving larger protein crystals.** *J Appl Cryst* 1997, **30**:198-202.

The use of oil in vapor diffusion trials.

Lab Exercises with Oils

Exercise 16.1. Setting up microbatch trials

Materials required:

1. Microbatch plates, also known as HLA, Terasaki, or microtiter plates, available from Hampton Research, Douglas Instruments, Nunc, Molecular Dimensions, etc.

2. Low density oils:
   - Paraffin oil—paraffin liquid colorless, light GPR (ρ = 0.83-0.86 g cm^{-3}) BDH cat. no. 29436 5H
   - Silicone oil—Dow Corning 200/1 cS silicone fluid (ρ = 0.82 g cm^{-3}) BDH cat. no. 63002 4N

   or
   - Al's oil—a mixture of the above oils from Hampton Research.

     (The Dow Corning oils are also available through other companies, e.g., Merck, ABCR, etc.)

Proteins and buffers required:

1. Lysozyme (Sigma cat. no. L 6876) at 40 mg/ml (40 mg lysozyme powder dissolved in 1 ml of 10 mM sodium acetate buffer, pH 4.7)

2. Thaumatin (Sigma cat. no. T 7638) at 100 mg/ml (100 mg dissolved in 1 ml of deionized water).

3. 12% NaCl in water

4. 1 M PIPES or ADA buffer pH 6.8

5. 2 M Sodium potassium tartrate

Procedure:

1. Pipette 5 ml of paraffin oil into a microbatch plate. The oil will spread over the plate and cover the wells.

2. Using a Gilson P2 pipette, withdraw 1 μl of a screening solution (from a screening kit or a "homemade" screen).

3. Insert the tip into the well under the surface of the oil (as shown in Figure 16.1a) and dispense the 1 μl drop. If you find it difficult to hold the tip in mid-oil, you can rest the edge of the tip on the floor of the plate as you dispense. As you withdraw the tip from the oil, the drop will detach from it and fall to the bottom of the well.

4. Now add in the same way 1 μl of protein solution to that well and mix gently with the pipette tip. The two (separate) 1 μl drops join and become a 2 μl drop.

Perform the identical exercise in several microbatch plates using 5 ml of different oils such as Al's oil, silicone oil, and mixtures of paraffin and silicone oils in different ratios. Observe the different speeds at which precipitate/crystals appear.

Examples with lysozyme and thaumatin:

1. Take 20 μl of lysozyme solution at 40 mg/ml and mix with 20 μl of 12% NaCl solution in a microcentrifuge tube.

2. Dispense drops of volumes ranging from 1-8 μl under the oil. Crystals will appear over 2 to 7 days. See if the quality of crystals varies as a function of drop size.

The wells of microbatch plates can contain maximum 10 μl drops. If you wish to dispense larger drops, use a different container (e.g., Linbro plate or Bijou vial); you will need at least 1-2 ml of oil per each trial to cover the drops in order to prevent evaporation.

Try the same procedure using equal volumes of 25-30 mg/ml thaumatin solution and 0.5-0.6 M NaK tartrate in 0.1 M PIPES or ADA buffer, pH 6.8. Crystals should appear overnight.

Exercise 16.2. Filtration experiments

Materials required:

1. Centrifugal filters: 0.2 μm and 300,000 molecular weight cutoff (MWCO). These are available from many sources, e.g., Millipore cat. no. UFC3 OGV 25 (0.22 μm) and Millipore cat. no. UFC3 TMK 25 (300 kD MWCO). We have found the Millipore filters to be the easiest and quickest for performing the exercises below. A packet containing a selection of Millipore filter sizes can be obtained from Molecular Dimensions.

2. Bench centrifuge.

Procedure:

1. Mix 150 μl of freshly made lysozyme solution with 150 μl of NaCl in a microcentrifuge tube as described in Exercise 16.1.

2. Divide the mixture into 3 aliquots (of 100 μl each):

   (a) Leave one aliquot unfiltered.

   (b) Filter the second aliquot through a 0.22 μm filter.

   (c) Filter the third aliquot through a 300,000 MWCO filter.

   Filtration is performed by placing the filters in a bench centrifuge at 2-9 g for 1-2 minutes.

3. Dispense droplets of each aliquot for crystallization under the oil.

   Expected results: You should get a number of crystals in the unfiltered solution, a smaller number of larger crystals in the solution which was filtered with a 0.22 μm filter, and no crystals at all in the solution which was filtered through the 300,000 MWCO filter.

4. Add a nucleant or seed to a trial which has been filtered through the 300,000 MWCO filter—see if you get crystals.

5. Repeat the identical experiments (a) using a lysozyme solution which has been stored in the cold room for over a week (b) using higher concentrations of lysozyme or NaCl (c) using the recipe for thaumatin. Observe the different results.

Exercise 16.3. Containerless crystallization

Materials required

1. Any vessel which can contain volumes between 0.3 and 10 ml is suitable: 7 ml Bijou containers, spectrophotometer cuvettes, Linbro plates, or microwell strips.

2. Oils
   - **Low density oil:** Dow Corning 200/5 cS silicone fluid ($\rho = 0.92$ g cm^{-3}) BDH cat. no. 63003 4P
   - **High density oil:** Dow Corning FS 1265/10,000 cS ($\rho = 1.27$ g cm-3) BDH cat. no. 63026 2C

   or
   - Containerless Crystallization Kit, Hampton Reserach cat. no. HR3-419. This kit comes complete with all necessary oils and vessels.

Procedure:

1. Place a layer of the high density oil so that it covers the floor of the vessel. The oil is very viscous and it may be easier to layer it using a spatula rather than a pipette.

2. Now pipette a thick layer of the low density oil on top of the high density oil (150 µl in microwell strips and >1 ml if using larger vessels). The two oils are immiscible and a flat meniscus will form at the interface (Figure 16.2, c).

3. Mix the crystallization trials in microcentrifuge tubes as described above or, if you have very little material, mix the crystallization drop on a cover slip or piece of parafilm.

4. Pipette the crystallization drops under the surface of the top layer of oil at the center of the container. The drop will situate itself at the interface of the two oils (Figure 16.2,c).

5. Seal your container with tape in the case of microwell strips or a lid in the case of other vessels.

TIP: Drops which are smaller than 5 µl are more likely to migrate to the walls of the container than larger drops. Paraffin oil cannot be used as the top layer since it forms a rounded meniscus which also causes the drops to move to the container walls. The low density oil needs to be slightly viscous (5 cS) in order to ensure that the drops do not migrate.

Exercise 16.4. Insertion of oil barrier to slow down vapor diffusion experiments

1. Mix paraffin and silicone oils in different ratios (equal volumes, 65% paraffin + 35% silicone, etc.). The oils are totally miscible. Shake well and let stand for several minutes until the bubbles disappear.

2. Set up your vapor diffusion experiment as you normally do, using the conditions (pH, concentrations, etc.) which give your best crystals.

3. Just prior to sealing the system over the reservoir, pipette a measured volume (e.g., 500 µl) of a mixture of paraffin and silicone over the reservoir. The oil will form a layer above the reservoir (Figure 16.3).

4. Seal the reservoir as usual (tape or grease).

Expected Results:

Lysozyme and thaumatin experiments (with no oil barrier) will produce crystals within 2 days. Trials containing an oil barrier can take up to 10 days to produce crystals. The speed of crystallization becomes slower when the thickness of the oil layer increases or the paraffin to silicone ratio is increased. Fewer, larger crystals are expected to grow over the longer periods compared to those grown in the control.

Variations:

1. Vary the thickness of your oil barrier by pipetting different volumes (from 100 µl to 1 ml) over the reservoir. Observe the difference in crystal quality.

2. Start by using a mixture of equal volumes of paraffin and silicone, then vary the ratio of paraffin to silicone.

TIP: This method has worked very well in cases where the precipitant is salt. It also works well with 2-12.5% PEG (various molecular weights) and MPD. Our laboratory is currently testing the method for higher (>12%) concentrations of PEG 4000 to 8000 and MPD.

17

Crystallization for Cryo-Data Collection

Elspeth F. Garman

17

Crystallization for Cryo-Data Collection

Elspeth F. Garman

University of Oxford, Oxford, U.K.

17.1. Introduction

The technique of cryo-crystallography, collecting X-ray data from crystals held at a temperature of around 100K, is becoming an essential tool of the macromolecular structural biologist. Radiation damage to crystals held at cryo-temperatures is negligible on a rotating anode X-ray generator, and substantially reduced at a synchrotron. This factor usually allows a complete data set to be collected from a single crystal, and results in generally higher quality and higher resolution diffraction data, giving in turn more accurate structural information. Although the reduction in radiation damage has been the driving force behind recent developments in cryo-methods, there are further advantages of low temperature data collection:

1. The mounting techniques are much gentler, causing little mechanical stress; this is beneficial if crystals are fragile.
2. Crystals can be screened and stored while in peak condition for later use.
3. Small crystals can be used more easily.

Detailed descriptions of the techniques and the ideas behind them can be found in references.[1,2]

The most widely used cryo-mounting method is to suspend the crystal in a film of an "antifreeze" solution (a "cryo-protected" solution in which the crystal is stable) held by surface tension across a small diameter loop[3] of fiber and quickly plunge it into a gaseous nitrogen stream or liquid cryogen. The fiber loops can be made by the experimenter[1,2] or purchased from Hampton Research.

The technique requires that the protein crystal be soaked in a solution which, when frozen, will form vitrified water (an amorphous glass) rather than crystalline ice, a so-called "cryo-protectant" solution. In practice this means the addition of a cryo-protectant agent such as glycerol, ethylene glycol, MPD, low molecular weight PEG (400 or 600), or sugar to the mother liquor or stabilizing buffer. The crystal is then immersed in this solution for anything between 0.5 seconds and 24 hours prior to being speedily flash-frozen. The effect of these agents is to lower the freezing point of the mother liquor, raise the temperature at which vitrification can be achieved, and allow more time for the freezing process to occur without ordered ice formation.

Why include a chapter on cryo-crystallography in this book? If the crystal has been grown in the presence of even a low concentration of a cryo-protecting agent, it is more likely that this concentration can be increased to the required value without damaging the crystal. Thus, it is well worth thinking about cryo-protection at the stage of setting up crystallization trials, and screening some conditions which include a cryo-agent (even if at low concentration e.g. 5-10% glycerol, PEG 400, or MPD). This makes flash-freezing more straightforward when you start to collect X-ray data. There have been cases where cryo-protection without damage to the crystal could only be achieved by co-crystallization with a cryo-agent.

Some crystallization conditions are already adequately cryo-protected: solutions 1, 5, and 13 of Hampton Crystal Screen (see reference[4] and appendix A 17.1 of this chapter) and the crystal can be flash-frozen straight from the mother liquor. This is the ideal situation, since it minimizes the handling of the crystal, resulting in lower mosaic spread and higher quality data. In fact, several crystal screens have been developed which require no additional cryo-protection prior to flash-freezing. These are commercially available from Emerald BioStructures, Inc., under the trade names Cryo I (48 solutions) and Cryo II (48 solutions) (U.S. Patents Pending).[5] (See Appendix A5 of this book.)

An alternative and increasingly used approach is to first determine crystallization conditions for your protein as per the rest of this book, find cryo-conditions for the successful mother liquor, and then to try crystallizing the protein in the cryo-protected mother liquor. This strategy is summarized in Figure 17.1 and given as a step-by-step protocol below.

17.2. Protocol for Finding Crystallization Conditions for Cryo-Crystallography

1. Obtain crystals of your protein by the usual coarse screening methods and optimize these conditions (pH, precipitate concentration, additives, etc.).

2. Check that the crystals diffract at room temperature by mounting one in a capillary tube (see Section 17.4). Note that if you omit this step and then the crystals do not diffract when flash-frozen, you will never know whether they have been damaged by the flash-freezing protocol or if they were of no use anyway.

3. If the crystals diffract at room temperature, it is time to find a suitable cryo-protectant. Make up stock solutions of double the strength of whatever is in the drops that yielded crystals. For example, if your crystals grew in 20% PEG 4000, 10mM Tris, make a stock solution of 40% PEG 4000, 20 mM Tris (2X stock solution). Now make up 5%, 10%, 15%, 20%, 25%, 30%, and 35% cryo-protectant in this 2X stock solution. For example, to make 5% cryosolution, add 5 μl 100% cryo-agent to 50 μl 2X stock solution and 45 μl H_2O. Thus the original concentration of your mother liquor is retained and is not diluted. In fact it is worth actually increasing the concentration of the original mother liquor by 2 to 4% as is usual for stabilizing solutions, (52 μl mother liquor and 43 μl water in the above example).

4. Which cryo-agent? For crystals grown in high salt concentration, try glycerol first. For crystals in low salt concentration, MPD or ethylene glycol may be successful. For crystals in PEGs ≥ 2000, try addition of PEG 400 or 600. For mother liquors which contain

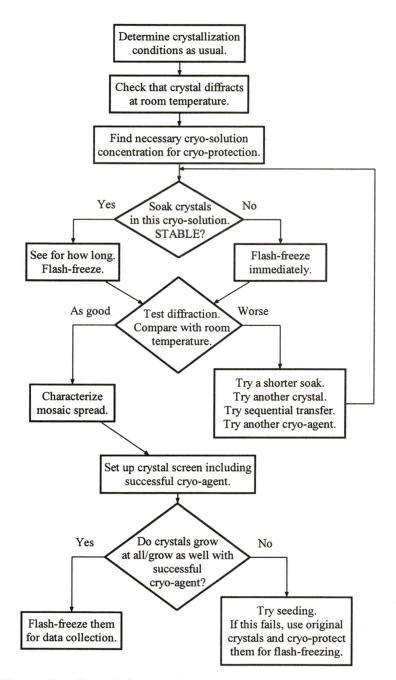

Figure 17.1. Strategy for crystallization in cryo-protected mother liquors.

MPD or PEGs with molecular weights under 2000, you can increase the concentrations of these. For most other cases, glycerol is a good place to start. For more exotic agents see references.[1,2]

5. Make sure the cryo-cooler on your X-ray equipment is working properly and is aligned on the crystal position. Some training in cryo-cooler operation is required before using one in an experiment.

6. Center your fiber loop at the crystal position using whatever mounting pin arrangement your laboratory has for cryo-data collection. See references.[1,2]

7. Place around 20 μl of the 5% cryo-solution in a shallow depression microscope slide or in a microbridge sitting in the well of a Linbro tray. Under the microscope, dip the loop in the cryo-solution so that a drop of similar volume to your crystal is held across the loop. Flash-freeze it as swiftly as possible by covering the nitrogen stream with a strip of cardboard (held by a second person if this is easier) and placing the loop and mounting pin on the goniometer (self-opposing angle-ended tweezers are a convenient tool for this) and then swiftly removing the cardboard. Dehydration of the drop can result in crystallization of mother liquor components, and this must be avoided.

8. Inspect the loop. If the frozen drop is cloudy (composed of ice crystals or crystallites) rather than glassy (vitrified), increase the cryo-agent concentration by 10% for the next trial. When the frozen drop is glassy, collect a diffraction image and check it for sharp (at 3.44Å, 3.67Å, 3.90Å) or diffuse ice rings. If these are present, increase the cryo-agent by 5% and repeat the diffraction experiment. Aim for a concentration of cryo-agent which gives a faint diffuse solvent scattering ring with a similar gradient slope on the high and low resolution sides. See Plate 25 and reference.[4]

9. Try transferring a crystal into this cryo-solution and watch it under the microscope for signs of crystal damage. If you observe any, flash-freeze it immediately. To do so, immerse the loop so that one edge is beneath the crystal and with the plane of the loop perpendicular to the liquid surface, gently pull it vertically upwards to

minimize the thickness of the film of liquid (see Figure 17.2). Flash-freeze it as in step 7 above. If there is no damage leave the crystal longer to see what happens. Immersion times between 0.5 sec and 24 hours have been used successfully but a time around 3 minutes is often appropriate. Now repeat the experiment with a new crystal and remove it from the cryo-solution before the time you observed damage last time, even if this is less than 1 second.

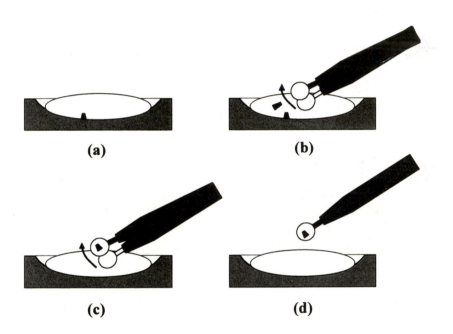

(a)   (b)

(c)   (d)

Figure 17.2. Steps to capture a crystal in a fiber loop for flash-freezing.

10. Center the crystal, taking care not to move it out of the nitrogen stream.

11. Test the diffraction and compare it to that obtained at room temperature. If it is not as good, try another crystal. Also try sequential soaking in increasing concentrations of cryo-solution, by

leaving the crystal in the microbridge and pipetting the cryo-solution onto it, agitating it, and drawing off a similar volume of liquid. This reduces the osmotic shock to the crystal. If all these fail, repeat the experiments with an alternative cryo-agent. Note that flash-freezing is, as yet, a rather empirical experimental technique for which there are few general rules regarding treatment of crystals and the optimum cryo-protectant conditions. An investment of time and crystals is required to determine them.

12. If possible, collect data and assess the diffraction quality, especially the mosaic spread. "Small" crystals (surface/volume ratio > 12 mm^{-1}) are usually better, giving lower mosaic spread. They are more mechanically robust, and cool faster due to their smaller thermal mass.

13. Once you have established satisfactory cryo-conditions, set up a new crystallization screen around refined conditions with and without the cryo-agent at the appropriate concentration. If this screen is unsuccessful, do not despair. The crystals can be flash-frozen as before, after a brief soak in the cryo-solution.

17.3. Laboratory Exercise for Determining Cryo-Protectant Solutions

The purpose of this experiment, to be performed without a crystal, is to show you the X-ray diffraction patterns from frozen solutions with different degrees of cryo-protection so that you can identify suitable cryo-protectant solutions which form vitreous ice when flash-frozen. You will need access to an X-ray generator and detector for this exercise.

1. Make up solutions of water/glycerol as follows: 100%/0%, 95%/5%, 90%/10%, 80%/20%, 70%/30%, 65%/35%, 60%/40%. Glycerol is easier to pipette accurately if you heat it in a water bath first to reduce the viscosity.

2. Follow steps 5 to 8 above. 60%/40% gives good cryo-protection and is the sort of image for which you should aim. Note that it is better than 65%/35% because the low and high resolution edges of

the diffuse scattering ring have more similar gradients. See Plate 25 for diffraction images from all the water/glycerol mixtures above.

17.4. Mounting Crystals for Room Temperature Data Collection

The main objective is to get the crystal into a mounting tube and seal the tube at both ends with wax without allowing the crystal to dry out or damaging it in the process. As protein crystals contain anything between 29% (edstin) to 95% (tropomyosin) solvent, if they dry out they shrink. Usually this decreases their three-dimensional order, and thus their diffraction quality. However, the crystal must be dry enough to prevent it moving during data collection, since even a small slippage can make a data set unprocessable. Some crystals are extremely fragile and difficult to mount, while others will withstand an amazing amount of handling. Below are some basic guidelines.

17.4.1. General Procedure for Mounting a Crystal in a Capillary

1. First find out if there are any limitations set by the data collecting equipment on the length of crystal mounting tube and position of crystal in tube, such as how much translation is available on the goniostat along the direction of the mounting tube.

2. Assemble all the equipment you need before opening up your crystallization tray, as the quicker you mount your crystal, the less time it has to dehydrate. You will need: some thin walled glass or fused quartz capillary tubes for the crystals, a syringe or a Gilson pipette, tips, a shallow dish with a greased lid or Linbro tray with cover slips to make it air tight, pasteur pipettes and a suction bulb, some drawn glass tubes (made from pulling Pasteur pipettes over a flame), tweezers, fine sewing needles, hypodermic needles, some thin slivers of filter paper, a soldering iron, some wax, strong nerves, and a steady hand.

3. Break the narrow sealed end of a capillary tube off with tweezers as cleanly as you can. A diamond knife can be used to scratch quartz tubes, and they can then be broken by hand. Glass tubes will break much more cleanly if a band of melted wax is put on them, and they are pinched with tweezers just beyond the wax. (Pinch off the wax, too. If your crystal lurks under the waxed part later on, you may think you have lost it.) Alternatively a hot wire can be used to cut the tube by melting it.

4. Load the tube onto the syringe or a Gilson pipette using a piece of flexible plastic hose. This flexibility means you will not break the mounting tube as often when extracting crystals from their drops.

5. If the crystals were grown in a microbridge or Linbro tray, they can be mounted straight from this. They may have to be transferred into a harvesting or stabilizing solution before mounting, in which case a microbridge standing in the solution in an otherwise empty tray can be a convenient holder. You can then move the crystals around easily under the microscope, and cover them after extracting the one you want, to prevent the remaining crystals from drying out.

6. Move the chosen crystal to the edge of the drop, using an acupuncture or sewing needle (hypodermic needles are unsatisfactory as they suck up mother liquor by capillary action and can thus beach the crystals) or other manipulator (e.g., cat hair or eyelash). Make sure the crystal is no longer stuck to the cover slip or in the "skin" on the surface of the drop. Try to move the crystal by agitating the liquid around it rather than actually touching it.

7. Gently suck up one crystal into the capillary, using one hand to steady the tube (easier with part of the hand resting on a surface) and one to operate the suction. (See Figure 17.3, a.)

8. Seal the open end of the tube with several thin layers of melted wax, letting each layer solidify before applying the next. Most soldering irons are much hotter than is necessary and vaporize a lot of wax. A succession of thin layers of wax, rather than great single blobs, minimizes the heat reaching the crystal.

9. Detach the tube from the syringe and place it upright in some plasticene to let the crystal fall towards the sealed end under its own weight. (See Figure 17.3, b.)

10. Remove the excess mother liquor from the tube using thin slivers of filter paper or the capillary action of narrow drawn glass tubing. (See Figure 17.3, c.) You will have to experiment with how wet/dry your crystal needs to be to survive.

11. Put a band of wax below the broad end of the capillary and break off the tube above it.

12. Put a small amount of mother liquor in the open end using the syringe and hypodermic needle, and then seal it with wax. (See Figure 17.3, d.)

13. Under the microscope, gently mark the tube with a felt tip pen on each side of the crystal position to make the crystal easier to locate.

14. If, in spite of your best drying efforts, your crystals still slip, leave your mounted crystal in the same orientation as you will use for data collection to settle for a day or two before starting. Tapered tubes are also available. Alternatively, use a flattened mounting tube to increase the contact area, and hence the adhesion, of the tube to the crystal. One side of a quartz capillary (glass is too fragile and disintegrates) can be flattened by placing the tube on a metal block which has semicircular slots of the same diameter as the tube, holding one end in place with a strip of plasticene (do not clamp both ends as this puts too much stress on the tube as it cools) and heating the tube with an oxy-acetylene flame. The top of the tube collapses to a flat surface. This is a difficult technique to get right and requires some practice.

15. Monitor the temperature inside the radiation housing of your data collection instrument, even for "room temperature" data collection. Many protein crystals cannot tolerate changes in temperature. For data collection at lower than room temperature in capillaries, a column of silicon oil can be placed on each side of the crystal after the excess mother liquor has been removed to minimize distillation in the tube due to temperature gradients. Also take care to minimize draughts around your crystal, again to avoid distillation of mother liquor in the tube.

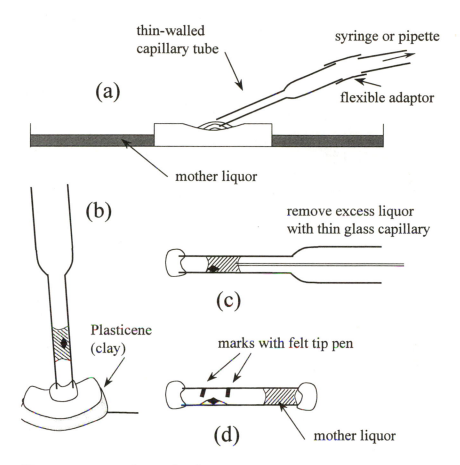

Figure 17.3. Crystal mounting for room temperature data collection.

References and Further Reading

1. Garman EF, Schneider TS: **Macromolecular cryocrystallography.** *J Appl Cryst* 1997, **30**:211-237.

    Comprehensive review of cryocrystallography, with particular emphasis on the practical aspects of the technique.

2. Rodgers, D: **Practical cryocrystallography.** *Meth Enzymol* 1997, **276**:183-203.

Detailed review of current experimental practice.

3. Teng T-Y: **Mounting of crystals for macromolecular crystallography in a free-standing thin film.** *J Appl Cryst* 1990, **23**:387-391.

First description of cryo-loop mounting method.

4. Garman EF, Mitchell EP: **Glycerol concentrations required for cryo-protection of 50 typical protein crystallization solutions.** *J Appl Cryst* 1996, **29**:584-587.

Glycerol concentrations required to cryoprotect the Hampton Screen I solutions.

5. The Cryo Matrices were developed by Steve L. Sarfaty and Wim G.J. Hol at the University of Washington (Seattle, WA, USA, 1998). Emerald BioStructures, Inc. has obtained an exclusive license from the University of Washington to market the Cryo I and II (U.S. Patents Pending).

A 17.1. Glycerol concentrations required to cryo-protect Hampton Crystal Screen I solutions.[4]

Courtesy of the IUCr.[4]

Minimum percentage of 100% glycerol to be added to solutions 1 to 50 of the Hampton Crystal Screen I (™) reagent components to provide cryo-protection when frozen to 100K. It must be noted that glycerol was added to the Crystal Screen solutions resulting in a **dilution** of the original components. (These solutions are now available from Hampton under the name "Crystal Screen Cryo," made up to the right concentrations for crystals grown in the Crystal Screen I.)

Solution number	SALT	BUFFER	PRECIPITANT	GLYCEROL % CONC. v/v
1	0.02 M Ca chloride	0.1 M Na acetate pH 4.6	30% v/v MPD	0
2	none	none	0.4 M K, Na tartrate	35
3	none	none	0.4 M NH$_4$ phosphate	35
4	none	0.1 M Tris HCl pH 8.5	2.0 M NH$_4$ sulfate	25
5	0.2 M Na citrate	0.1 M Na HEPES pH 7.5	30% v/v MPD	0
6	0.2 M Mg chloride	0.1 M Tris HCl pH 8.5	30% w/v PEG 4000	20
7	none	0.1 M Na cacodylate pH 6.5	1.4 M Na acetate	30
8	0.2 M Na citrate	0.1 M Na cacodylate pH 6.5	30% v/v 2-propanol	30

Continued on next page

Solution number	SALT	BUFFER	PRECIPITANT	GLYCEROL % CONC. v/v
9	0.2 M NH$_4$ acetate	0.1 M Na citrate pH 5.6	30% w/v PEG 4000	15
10	0.2 M NH$_4$ acetate	0.1 M Na acetate pH 4.6	30% w/v PEG 4000	15
11	none	0.1 M Na citrate pH 5.6	1.0 M NH$_4$ phosphate	30
12	0.2 M Mg chloride	0.1 M Na HEPES pH 7.5	30% v/v 2-propanol	10
13	0.2 M Na citrate	0.1 M Tris HCl pH 8.5	30% v/v PEG 400	0
14	0.2 M Ca chloride	0.1 M Na HEPES pH 7.5	28% v/v PEG 400	5
15	0.2 M NH$_4$ sulfate	0.1 M Na cacodylate pH 6.5	30% w/v PEG 8000	15
16	none	0.1 M Na HEPES pH 7.5	1.5 M Li sulfate	25
17	0.2 M Li sulfate	0.1 M Tris HCl pH 8.5	30% PEG 4000	15
18	0.2 M Mg acetate	0.1 M Na cacodylate pH 6.5	20% PEG 8000	20
19	0.2 M NH$_4$ acetate	0.1 M Tris HCl pH 8.5	30% v/v 2-propanol	20
20	0.2 M NH$_4$ sulfate	0.1 M Na acetate pH 4.6	25% w/v PEG 4000	20
21	0.2 M Mg acetate	0.1 M Na cacodylate pH 6.5	30% v/v MPD	0
22	0.2 M Na acetate	0.1 M Tris HCl pH 8.5	30% w/v PEG 4000	15
23	0.2 M Mg chloride	0.1 M Na HEPES pH 7.5	30% v/v PEG 400	0
24	0.2 M Ca chloride	0.1 M Na acetate pH 4.6	20% v/v 2-propanol	30
25	none	0.1 M imidazole pH 6.5	1.0 M Na acetate	30
26	0.2 M NH$_4$ acetate	0.1 M Na citrate pH 5.6	30% v/v MPD	0
27	0.2 M Na citrate	0.1 M Na HEPES pH 7.5	20% v/v 2-propanol	30
28	0.2 M Na acetate	0.1 M Na cacodylate pH 6.5	30% w/v PEG 8000	15
29	none	0.1 M Na HEPES pH 7.5	0.8 M K,Na tartrate	35
30	0.2 M NH$_4$ sulfate	none	30% w/v PEG 8000	15
31	0.2 M NH$_4$ sulfate	none	30% w/v PEG 4000	15
32	none	none	2.0 M NH$_4$ sulfate	25
33	none	none	4.0 M Na formate	10
34	none	0.1 M Na acetate pH 4.6	2.0 M Na formate	30
35	none	0.1 M Na HEPES pH 7.5	1.6 M Na, K phosphate	25
36	none	0.1 M Tris HCl pH 8.5	8% w/v PEG 8000	35
37	none	0.1 M Na acetate pH 4.6	8% w/v PEG 4000	30
38	none	0.1 M Na HEPES pH 7.5	1.4 M Na citrate	10
39	none	0.1 M Na HEPES pH 7.5	2% v/v PEG-400 and 2.0 NH$_4$ sulfate	15
40	none	0.1 M Na citrate pH 5.6	20% v/v 2-propanol and 20% w/v PEG 4000	5
41	none	0.1 M Na HEPES pH 7.5	10% v/v 2-propanol and 20% w/v PEG 4000	15
42	0.05 M K phosphate	none	20% w/v PEG 8000	20
43	none	none	30% v/v PEG 1500	20
44	none	none	0.2 M Mg formate	50
45	0.2 M Zn acetate	0.1 M Na cacodylate pH 6.5	18% w/v PEG 8000	20
46	0.2 M Ca acetate	0.1 M Na cacodylate pH 6.5	18% w/v PEG 8000	20
47	none	0.1 M Na acetate pH 4.6	2.0 M NH$_4$ sulfate	20
48	none	0.1 M Tris HCl pH 8.5	2.0 M NH$_4$ phosphate	20
49	1.0 M Li sulfate	none	2% w/v PEG 8000	20
50	0.5 M Li sulfate	none	15% w/v PEG 8000	20

18

Crystallization of Membrane Proteins

Jeff Abramson and So Iwata

18

Crystallization of Membrane Protein

Jeff Abramson and So Iwata

Uppsala University, Uppsala, Sweden

Crystallization of soluble proteins has become relatively routine with over 6,000 three-dimensional structures determined by X-ray crystallography. In contrast, there are a mere handful of identified integral membrane protein structures. This disproportion is not due to lack of effort; rather, it lies in the difficulties encountered in the crystallization of membrane proteins. Nevertheless, there are still some rational approaches which, when applied, increase the probability of obtaining good membrane protein crystals. It is essential, however, to understand some basic principles used for membrane protein crystallization. This chapter summarizes these principles and then presents a practical portion which will show some examples and techniques used by membrane protein crystallographers.

18.1. Principles of Membrane Protein Crystallization

Membrane proteins are embedded within the lipid bilayer and contain both a hydrophilic surface exposed to the solvent and a hydrophobic region buried within the lipid bilayer (Figure 18.1). To solubilize the membrane protein from the lipid bilayer, a so-called "mild detergent" is added. The detergent mimics the lipid bilayer by covering the hydrophobic region of the protein, generating a protein-detergent micelle. Once the membrane protein has been successfully solubilized and purified, crystallization trials may proceed.

Membrane protein crystals are categorized in three ways depending on how the crystals are formed. Of these, 2D crystals are, in principle, reconstituted biomembranes used in electron microscopy and will not be discussed in this chapter. Instead, we will focus on 3D crystals which are used in X-ray crystallography. These are termed type I and type II crystals.

Type I crystals are stacked 2D crystals. These crystals are stabilized through hydrophobic interactions of the reconstituted membrane and polar interactions between each stacked reconstituted membrane. Due to the difficulties in reinforcing both hydrophobic and polar interactions during the crystallization process, it is hard to obtain large type I crystals. Recently, however, a lipidic cubic phase was reported to be useful for generating type I crystals.[1] A lipidic cubic phase is a three-dimensional continuous lipid phase which acts like a solvent for membrane proteins. The detailed mechanism by which type I crystals form in this phase is unknown. So far, only the structure of bacteriorhodopsin has been solved using this technique, but it has a promising future in the field.

The most common type of membrane protein crystals obtained, as well as attempted, are type II crystals. Type II crystals are grown using the conventional precipitants (ammonium sulfate, PEG, etc.) and techniques applied to ordinary soluble proteins, with the difference that a detergent must be present.

However, the influences from detergents have two impeding characteristics not present in soluble protein crystals. The first is that protein-protein contacts, which are essential for the formation of the crystal lattice, can only be mediated through the hydrophilic surfaces of the membrane protein. The detergent molecules covering the hydrophobic region prevent any specific protein-protein contacts. A method which extends the hydrophilic surface of the protein using an antibody fragment has recently been developed.[2] Increasing the hydrophilic surface facilitates the formation of these essential protein-protein contacts. So far there has been only one successful example utilizing this method, cytochrome c oxidase from *Paracoccus denitrificans*. In the near future, however, this will likely be a common strategy for acquiring membrane protein crystals.

The second impediment stems from the filling of the space between proteins with detergent micelles. The detergent micelle must fit perfectly in this open space between proteins. Since the size of this space is unknown before solving the structure, the optimal detergent is found by trial and error. A schematic representaion of solubilization, purification, and the different kinds of membrane crystals can be seen in Figure. 18.1.

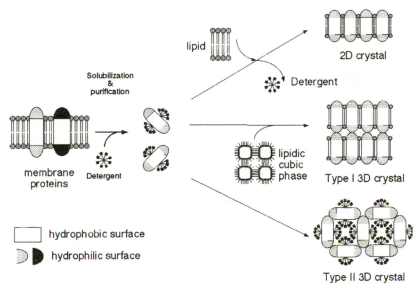

Figure 18.1. Types of membrane protein crystals

18.2. Practical Approach for Membrane Protein Crystallization

The crystallization of membrane proteins most often proceeds in the same manner as for soluble proteins, except for the existence of detergent in the crystallization conditions. However, the detergent provides an additional dimension to membrane protein crystallization trials which frequently prevents acquiring good crystals. In this portion of the chapter, we summarize the most critical parameters for membrane protein crystallization. For a more detailed discussion, refer to references.[3,4]

18.3. Sample Preparation

The existence of a pure and homogenous sample at the proper concentration is a prerequisite for all crystallization trials. The amphiphilic nature of membrane proteins requires the use of detergents, which leads to many unique problems not found with the preparation of soluble proteins. We summarize them in this section.

18.3.1. Selection of Detergent

The detergent used for purifying a membrane protein is normally different from the one applied in the crystallization trial. A good dispersing detergent is one which has a low critical micellar concentration (CMC), is inexpensive, free from contaminants, and stabilizes the protein to permit its purification. Detergents like 0.03% dodecyl maltoside or 0.2% LDAO (N,N-dimethyldodecylamine-N-oxide) are good choices. (CMC is explained in section 18.4.1.)

18.3.2. Purification

Essentially, separation methods for membrane and soluble proteins are the same. Many purifications proceed through various chromatographic techniques. However, membrane proteins tend to have strong interactions with the chromatography media, bringing about a lower column efficiency. This makes affinity chromatography, when possible, an ideal choice for its capabilities of performing a complete separation in one step. Another problem arising in the purification process is over-purification, which is often disadvantageous. This is particularly dangerous for large membrane protein complexes, which can often lose weakly bound subunits. The crystals of the two largest protein complexes, cytochrome c oxidase and cytochrome bc_1 complex, have both been purified by the less aggressive iterative salt precipitation method. Due to these drawbacks in chromatography techniques, sedimentation and preparative iso-electric focusing are frequently used for membrane protein purifications. In order to maintain the integrity of the sample, the quality should be checked using the conventional PAGEs (SDS and native). Also, iso-electric focusing and analytical ultracentrifugation are other useful techniques for ascertaining the quality of the protein.

18.3.3. Protein Concentration and Detergent Exchange

Before crystallization, the protein should be solubilized at a reasonable concentration in a sensible buffer. Ideally, it is best to prepare samples in lower buffer concentrations (e.g., 5 mM) without salt. If the sample precipitates, then incrementally increase the salt (50, 100, 150 mM NaCl). In general 10 mg/ml is a good starting protein concentration. However, for large protein complexes, like cytochrome c oxidase and cytochrome bc1 complex, concentrations up to 50-80 mg/ml may be needed. It is, however, more suitable to look at concentration in terms of molarity when considering the size of the protein for crystallization. Aim for 100-200 µM.

18.4. Crystallization

As previously mentioned, crystallization trials for membrane proteins are similar to those for soluble proteins. Here, we will present some critical points unique to membrane protein crystallization.

18.4.1. Selection of the Detergents

The proper selection of detergent is the most critical parameter for membrane protein crystallization. Here are a few useful suggestions on where to begin.

1. It is essential that the protein is soluble and stable in the detergent. Often membrane proteins are "happier" in detergents with longer alkyl chains since these more closely mimic the membrane. However, detergents that form smaller micelles can leave more of the hydrophilic region exposed. Therefore, a compromise must be made between these two alternatives. The best way to begin is to try detergents that have been successfully used in previous membrane proteins crystals (Table 18.1).

2. Detergent molecules are monomerically distributed until they reach a threshold concentration where they spontaneously form micelles. This is known as the critical micellar concentration (CMC). It is in this micellar state that detergents are effective for solubilizing. The detergent concentration is usually kept 1-3 times greater than its CMC to insure solubilization. The CMC is inversely related to the size of the alkyl chain and hence larger detergents have lower CMCs. Precautions should be taken because the CMC is sensitive to both temperature and salt concentrations. The product catalog of Anatrace has a good list of CMCs for many detergents.[5]

3. Some detergents, including glucosides and maltosides, are slowly hydrolyzed in solution. It is therefore better to use freshly prepared detergent solutions. Detergent source is important. It is better not to change the brand of detergent once initial crystals have been obtained. A large selection of high-quality detergents are available from Calbiochem and Anatrace.[5]

4. A complete detergent exchange is often needed before the crystallization trial. This can be accomplished in different ways. Ultrafiltration and dialysis techniques are commonly used but are rather inefficient, particularly for larger micelles. When using these techniques make sure the membrane filter has a large enough pore size to permit replacing the detergent micelle. The most effective technique for changing detergent is by applying the sample to a small ion exchange column (e.g., 1 ml of Q-sepharose FF, Pharmacia) and slowly washing it, for at least five column volumes, with the crystallization buffer containing the target detergent.

18.4.2. Crystallization Setups

The most commonly performed setup is the vapor diffusion method. Of these, the sitting drop technique is more popular for membrane proteins than the hanging drop. Hanging drops tend to spread over the surface of the siliconized cover slip because of the decrease in surface tension caused by the detergent. However, hanging drops can be implemented when the drop size is small (<5 µl) and are useful for initial screenings.

Table 18.1. Crystallization conditions of membrane proteins

Protein	Source	Detergent	Precipitant	Buffer	Salt	Additive	Setup	Temp.	Ref.
Reaction center	*R. viridis*	LDAO 0.5%	AS 2.2–2.4 M	AS pH 6.0	-----	HT 3%	SD	RT	6
Light harvesting complex	*R. acidophila*	OG 1%	AS 2.2 M	P 0.9 M pH 9.5	-----	Benz-amidine 2.5%	VD	20°C	7
Light harvesting complex	*R. molischianim*	UDAO 0.2%	AS 3.0–3.3 M	PP pH 6.5	-----	HT 3.2%	SD	20°C	3
Reaction center	*R. sphaeroides*	LDAO 0.1%	PP 1 M	PP pH 6.5–7.0	-----	HT 3% Dioxane 1%	SD	RT	8
Photosystem I	*R. sphaeroides*	DDM 0.02%	Salt-in	MES 5 mM pH 6.4	$MgSO_4$ 0.1 M	-----	DI	---	9
Porin	*E. coli*	OHS 0.6% C_8E_4 0.1%	PEG 2000 10.5%	Tris 50 mM pH 9.8	$MgCl_2$ 0.7 M	-----	DI	RT	10
Porin	*R. capsulatus*	C_8E_4 0.6%	PEG 600 23–30%	Tris 20 mM pH ?	LiCl 0.3 M	-----	HD	20°C	11
Maltoporin	*E. coli*	DM 0.4% $C_{12}E_9$ 0.1%	PEG 2000 15–18%	HEPES 20 mM pH 7.0	$MgCl_2$ 0.1 M	-----	DI	RT	12

Table 18.1—*continued*

			AS 2.5 M mmPEG 5000 0.25%	AS pH 7.4	NaCl 75 mM	-----	HD	RT	13
α-Hemolysin	*S. aureus*	OG 25 mM							
K$^+$ channel	*S. lividans*	LDAO 5 mM	PEG 400 48%	Tris 50 mM pH7.5	KCl 0.15 M	DTT 2 mM	SD	20°C	14
Prostaglandin synthase	Sheep	OG 0.6%	PEG 4000 4–8%	SP 20 mM pH 6.7	NaCl 0.1–0.2M	-----	HD	-----	15
Cytochrome *c* oxidase	*P. denitrificans*	DDM 0.03%	mmPEG 2000 10–12%	AA 0.4 M pH 7.0	-----	DMSO 7%	SD	14°C	2
Cytochrome *c* oxidase	Bovine heart mitochondria	DM 0.2%	PEG 4000	SP 40 mM pH 6.8	-----	-----	B	4°C	16
Cytochrome *bc*$_1$ complex	Bovine heart mitochondria	DDM 0.15%	PEG 4000 6–8%	SP 50 mM pH 6.8	NaCl 100 mM	-----	SD	4°C	17
Cytochrome *bc*$_1$ complex	Bovine heart mitochondria	DMG 0.1 %	PEG 4000 12%	MOPS 50 mM pH 7.2	KCl 0.5M	Glycerol 20%	VD	0–4°C	18

Abbreviations:

DETERGENTS
LDAO	N,N-dimethyldodecylamine-N-oxide
UDAO	N,N-dimethylundecylamine-N-oxide
OG	octyl-β-D-glucopyranoside
DM	decyl-β-D-maltoside
DDM	dodecyl-β-D-maltoside
C$_{12}$E$_9$	dodecyl-β-nonaoxyethylene
DMG	decanoyl-N-methylglucamide
OHS	octyl-2-hydroxyethylsulfoxide
C$_8$E$_4$	octyltetraoxyethylene ether

BUFFERS/PRECIPITANTS
AS	Ammonium sulfate
P	Phosphate
PP	Potassium phosphate
SP	Sodium phosphate
AA	Ammonium acetate
PEG	Polyethylene glycol
mmPEG	PEG, monomethyl ether

OTHER
DI	Dialysis
SD	Sitting drop
VD	Vapor diffusion
B	Batch
RT	Room temperature
DTT	Dithiothreitol
DMSO	Dimethyl sulfoxide
HT	Heptane 1,2,3-triol

Hanging drops can often give different results than sitting drops and, therefore, it is a good idea to try both.

18.4.3. Precipitants

When using non-ionic detergents, the most successful precipitant, so far, is PEG (400-4000) with some salt (100 mM NaCl). For ionic detergents, the conditions often include ammonium sulfate and potassium phosphate. Membrane proteins tend to be unstable in high concentrations of organic solvents (such as MPD) and these are therefore avoided.

18.4.4. Temperature

The detergent micelle size is temperature sensitive. Crystallization trials should be performed at least at two temperatures, usually 20°C and 4°C. Most membrane protein crystals have been obtained at these temperatures. If possible, try another temperature between 10 and 15°C.

18.4.5. Screening Kits

For initial screening we usually use the hanging drop technique, where the protein and reservoir solutions are mixed in a 1:1 ratio. There is no need to keep detergent in the reservoir solution, so simply use ready-made kits utilizing the sparse matrix approach, (e.g., Crystal Screen I, Crystal Screen II, and MembFac from Hampton Research) in the same manner as soluble proteins. However, the PEG concentrations in the Crystal Screen (often 30%) are usually too high for membrane proteins. Therefore, we recommend a combination of Crystal Screen without the PEG conditions and the PEG conditions from MembFac for initial screening.

18.5. Optimization

The probability of getting membrane protein crystals on the first trial is less than a few percent. Once crystals are obtained, they will almost certainly need optimization. The most successful approaches for optimization are (1) by modifying the detergent micelle with additives and (2) using detergents with other size micelles. It has been reported that mild changes to the micelle can give dramatic difference in the crystal's quality and resolution. In order to ascertain the quality (resolution, mosaicity, anisotropy, radiation sensitivity, twinning, etc.) of the crystal, it must be properly characterized by X-ray diffraction. However, laboratory X-ray sources tend to be too weak for characterization, and synchrotron facilities should be used.

Additives are quite important for the optimization of, and in some cases initially obtaining, membrane protein crystals. Small amphiphile like heptane 1,2,3-triol and benzamidine are found to be useful additives, as these molecules can form mixed micelles with detergents and slightly alter the detergent micelle size.[3] Simultaneously, they prevent phase separations which can lead to protein denaturation. DMSO, dioxane, and some alcohols (e.g., ethanol, methanol, and MPD) have also been reported as effective additives for membrane proteins. Our laboratory often uses the additive screens from Hampton Research. Hampton Research also offers detergent screening kits which may be used to form mixed micelles and in the same manner generate slight differences in the micelle size.

References and Further Reading

1. Landau EM, Rosenbusch JP: **Lipidic cubic phases: a novel concept for the crystallization of membrane proteins.** *Proc Natl Acad Sci* 1996, **93**:14532-14535.

   The crystallization of bacteriorhodopsin within a lipid phase. The detergent diffuses into the lipid phase and crystal growth proceeds through lateral diffusion of the protein molecules.

2. Ostermier C, Iwata S, Ludwig B, Michel H: **F$_v$ fragment mediated crystallization of the membrane protein bacterial cytochrome c oxidase.** *Nat Struc Biol* 1995, **2**:842-846.

   Co-crystallization of cytochrome c oxidase with an antibody F$_v$ fragment.

3. Michel H: *Crystallization of membrane proteins.* Boca Raton, Florida: CRC Press; 1990.

   Detailed book on crystallizing membrane proteins.

4. Ostermier C, Hartmut M: **Crystallization of membrane proteins.** *Curr Opin Struct Biol* 1997, 7:697-701.

A review of membrane protein crystallization.

5. Product catalogue, Anatrace, Inc: 434 West Dussel Dr., Maumee, OH 43537 USA, telephone: (419) 891-3030. e-mail: anatrac434@aol.com.

Commercial catalogue for a detergent company. Contains useful information about the detergents.

6. Michel H: **Three-dimensional crystals of a membrane protein complex.** *J Mol Biol* 1982, **158**:567-572.

The photosynthetic reaction center is the first membrane protein structure.

7. Papiz MZ: **Crystallization and characterization of two crystal forms of the B800-850 light harvesting complex from *Rhodopseudomonas acidophila* strain 10050.** *J Mol Biol* 1989, **209**:833-835.

Crystallization conditions for two crystal forms of the light harvesting complex.

8. Buchanan SK, Fritzsch G, Ermler U, Michel H: **New crystal form of the photosynthetic reaction centre from *Rhodobacter sphaeroides* of improved diffraction quality.** *J Mol Biol* 1993, **230**:1311-1314.

Crystallization conditions for new crystal forms of the photosynthetic reaction center.

9. Krauss N, Hinrichs W, Witt I, Fromme P, Pritzkow W, Dauter Z, Betzel C, Wilson KS, Witt HT, Saenger W: **Three-dimensional structure of system I of photosynthesis at 6Å resolution.** *Nature* 1993, **361**:326-331.

The structure of system I of photosynthesis.

10. Pauptit RA, Zhang H, Rummel G, Schirmer T, Jansonius J, Rosenbusch JP: **Trigonal crystals of porin from *Escherichia coli.*** *J Mol Biol* 1991, **218**:505-507.

Crystallization conditions for trigonal crystals of porin.

11. Kreusch A, Weiss MS, Welte W, Weckesser J, Schulz GE: **Crystals of an integral membrane protein diffracting to 1.8 Å resolution.** *J Mol Biol* 1991, **217**:9-10.

The structure of porin.

12. Stauffer KA, Page MGP, Hardmeyer A, Keller TA, Pauptit RA: **Crystallization and preliminary X-ray characterization of maltoporin from *Escherichia coli.*** *J Mol Biol* 1990, **211**:297-299.

Crystallization conditions for maltoporin.

13. Gouaux JE, Braha O, Hobaugh MR, Song L, Cheley S, Shustak C, Bayley H: **Subunit stoichiometry of staphylococcal α-hemolysin in crystals and on membranes: a heptameric transmembrane pore.** *Proc Natl Acad Sci* 1994, **91**:12828-12831.

Crystallization conditions for α-hemolysin.

14. Doyle DA, Cabral JM, Pfuetaner RA, Kuo A, Gulbis JM, Cohen SL, Chait BT, MacKinnon R: **The structure of the potassium channel: molecular basis of K+ conduction and selectivity.** *Science* 1998, **280**:69-77.

The structure of K+ channel.

15. Picot D, Loll PJ, Garavito M: **The X-ray structure of the membrane protein prostaglandin H₂ synthase-1.** *Nature* 1994, **367**:243-249.

The structure of prostaglandin H₂ synthase-1.

16. Tsukihara T, Aoyama H, Yamashita E, Tomizaki T, Yamaguchi H, Shinzawa-Itoh K, Nakashima R, Yaono R, Yoshikawa S: **Structures of metal sites of oxidized bovine heart cytochrome c oxidase at 2.8Å.** *Science* 1995, **269**:1069-1074.

The structures of metal sites of bovine heart cytorome c oxidase.

17. Lee JW, Chan M, Law TV, Kwon HJ, Jap BK: **Preliminary cryocrystallographic study of the mitochondrial cytochrome bc₁ complex: improved crystallization and flash-cooling of a large membrane protein.** *J Mol Biol* 1995, **252**:15-19.

Crystallization conditions for the mytochondrial cytochrome bc₁ complex.

18. Yu C-A, Xia J-Z, Kauchurin AM, Yu L, Xia D, Kim H, Deisenhofer J: **Crystallization and preliminary structure of beef heart mitochondrial cytochrome bc₁ complex.** *J Mol Biol* 1996, **1275**:47-53.

Crystallization conditions for the mytochondrial cytochrome bc₁ complex.

19

One Last Piece of Advice: Appearances Can Be Deceiving!

Alex Cameron

19

One Last Piece of Advice: Appearances Can Be Deceiving!

Alex Cameron

Uppsala University, Uppsala, Sweden

It is every crystal grower's dream to look into their crystallization drops and find crystals like those in Plate 26, a or b. On the other hand, the crystals in Plate 26, c, would not be greeted with the same degree of enthusiasm. Looks, however, can be deceiving. The PDGF (platelet derived growth factor) crystals in Plate 26, a and b, diffract only to approximately 6Å. The best data set collected from glyoxalase I in Plate 26, c, extends to a resolution of 1.7Å. Crystals such as these, with dimensions 0.3 x 0.3 x 0.3 mm, were in fact used to solve the structure of the protein.[1] No evidence of the crystal shape was seen in the reflection profiles.

The moral of the story: It is the internal order of the crystal which is important in protein crystallography, not the outward appearance. Often the two go hand in hand, but a true assessment of crystal quality can only be made by X-ray diffraction.

References and Further Reading

1. Cameron A, Olin B, Ridderström M, Mannervik B, Jones TA: **Crystal structure of human glyoxalase I-evidence for gene duplication and 3D domain swapping.** *EMBO J* 1997, **16**:3386-3395.

   The excellent quality of the data collected from these ugly crystals can be seen in this paper.

A-Z Tips

A

Additives (*Naomi Chayen*)

There are several cases where anti-fungal/anti-microbial agents such as thymol, m-cresol,[1] and phenol[2] have proved to be crucial additives, for the growth and stability of crystals or for better ordered crystal forms.

These agents do not appear in the standard screens. Phenol can be used at 10-30 mM in the reservoir. Thymol is only soluble to 5 mM and effective as an additive at 2.5-3 mM. Alternatively, just add it as a solid grain into the reservoir. Thymol is volatile and needs to be replenished in the reservoir every 3-4 weeks.

References:

1. Chayen NE, Lloyd LF, Collyer CA, Blow DM: **Trigonal crystals of glucose isomerase require thymol for their growth and stability.** *J Cryst Growth* 1988, **97**:367-374.

2. Skarzynski T: **Crystal structure of α-dendrotoxin from the green mamba venom and its comparison with the structure of bovine pancreatic trypsin inhibitor.** *J Mol Biol* 1992, **224**:671-683.

Aging (*Naomi Chayen*)

Proteins can degrade during storage, due to denaturation, proteolysis, or other reasons. A protein solution stored at 4°C for a while may produce far worse quality crystals compared to the fresh protein.[1] In many cases, this problem can be overcome by filtering the protein immediately prior to setting up the crystallization trials. See *Filtration*.

Reference:

1. Chayen NE, Radcliffe JW, Blow DM: **Control of nucleation in the crystallization of lysozyme.** *Protein Science* 1993, **2**:113-118.

Azide (*Terese Bergfors*)

Buffers and PEG solutions can become contaminated by microbial growth during longer periods of storage. It is common practice to add sodium azide (NaN_3) to the solutions to prevent this. Some people add it to the protein solution as well. How much azide should be added? Different concentrations appear in the literature: 1 mM (=0.0065%), 0.02%, and 0.1%. A 0.02% concentration seems to be the most usual. Make a 20% stock solution of NaN_3, then add 1 ml per liter buffer.

Keep the following in mind if you use sodium azide:

1. It is toxic for people as well as microbes.
2. It is an inhibitor for some proteins (carbonic anhydrase and superoxide dismutase, to name two examples from our lab). You may be unintentionally adding a ligand to your protein.
3. It interferes with heavy atom derivatization.[1] Some metal azides are explosive.
4. There are examples where eliminating NaN_3 from the solutions improved the quality of the crystals.

An alternative to the use of NaN_3 (besides using other toxic bacteriocides like thymol or Thimerosal) is to filter your solutions through a 0.22 mm filter and store them at 4°C. Solutions should always be filtered even if azide is used.

Reference:

1. Petsko GA: **Preparation of heavy atom derivatives.** *Meth Enzymol* 1985, **114**:147-156.

B

Biological Macromolecule Crystallization Database (BMCD)
(*Terese Bergfors*)

This is a database of crystallization conditions for 1,465 biological macromolecules. See http://ibm4.carb.nist.gov:4400/.

Reference:

1   Gilliland GL, Bickham DM: **The biological macromolecule crystallization database: a tool for developing crystallization strategies.** *Methods: Companion to Meth Enzymol* 1990, **1**:6-11.

Buffers (*Enrico Stura*)

Buffer selection is an important consideration. Several biological buffers are based on sulfonates and act as phosphate mimics. HEPES was found in the active site of phosphotyrosyl phosphatase.[1] When optimizing crystallization conditions a change in buffer may not just change the pH. It may also involve a change of ligand.

When crystallizing a protein that needs calcium, zinc, or some other divalent metal, good buffer choices are cacodylate and acetate.

Reference:

1.   Zhang M, Zhou M, Van Etten RL, Stauffacher CV: **Crystal structure of bovine low molecular weight phosphotyrosyl phosphatase complexed with the transition state analog vanadate.** *Biochemistry* 1997, **36**:15-23.

C

Calcium (*Enrico Stura*)

Calcium ions have very poor solubility. Use acetate or cacodylate buffers. (See *Buffers* above.) Calcium ammonium phosphate is insoluble at even micromolar concentration. This salt is responsible in many drops for crystals which are not protein.

Centrifugation (*Naomi Chayen*)

It is common practice to centrifuge the protein solution immediately prior to setting up crystallization trials to sediment dust particles, precipitated protein, etc. This is usually done in a table centrifuge at 13,000 g for 1-2 minutes. When optimizing the conditions for crystallization it is a good idea to centrifuge your protein first, but I do not recommend it in the screening phase. It could remove a crucial heterogeneous nucleant that aids the crystallization of your sample.

See also *Filtration*.

Cover slips (*Terese Bergfors*)

Glass cover slips should be silanized. These are made in different thicknesses; use the ≥0.2 mm size. Plastic cover slips (Nunc) are also an option; they do not require silanization and are unbreakable.

Cysteines; protection of free cysteines;

see *Reducing agents*.

D

Detergents (*Terese Bergfors*)

Detergents are required in the crystallization of membrane proteins, but are a useful class of additives even for soluble proteins. This was first reported by McPherson for β-octylglucoside.[1] In fact, my first choice of optimization additive is 0.25% β-octylglucoside. Add the detergent to the drop (it does not have to be included in the reservoir). The concentration in the drop should be under the critical micelle concentration of the detergent.

When screening optimization additives, try at least several detergents as they have quite different effects. Don't try just β-octylglucoside and then think you have examined "detergents." Hampton Research is a good source of detergents for screening; they have three kits (a total of 72 detergents).

The pH, critical micelle concentration, conductivity, and absorbance at 275 nm vary for different brands of the same detergent and even from batch to batch. Buy the best grade of detergents possible or perform a quality control check on each new source. The detergents should be stored frozen after you have dissolved them.

Reference:

1. McPherson A, Koszelak S, Axelrod H, Day J, Robinson L, McGrath M, Williams R, Cascio D: **The effects of neutral detergents on the crystallization of soluble proteins.** *J Cryst Growth* 1986, **76**:547-553.

Drop; effect of varying the protein: reservoir volumes in the drop (*Madeleine Riès-Kautt*)

Vapor diffusion hanging drops are usually made by mixing equal volumes of the protein and reservoir solutions (e.g., 2 μl plus 2 μl). Thus, in a drop setup using a 10 mg/ml protein solution, the protein concentration will be initially 5 mg/ml and reach 10 mg/ml after equilibration against the reservoir.

By playing with the ratio of protein and reservoir volumes, the final concentration of protein in the drop can be varied.

Examples:
- 2 μl of 10 mg/ml protein plus 4 μl reservoir will give a final concentration of 5 mg/ml.
- 4 μl of 10 mg/ml protein plus 2 μl reservoir will give a final concentration of 20 mg/ml.

Dust (*Terese Bergfors*)

Dust on the cover slips or in the microbatch experiments acts as a nucleant. Usually its inclusion is unintentional. To insure clean drops, wipe each cover slip with a Kimwipe (or similar lint-free tissue). Dust can also be blown off with "canned air" of the type sold in photo shops and record stores.

Dyes; to test if crystals are salt or protein (*Elisabeth Sauer-Eriksson*)

IZIT (Hampton Research) and methylene green are two examples of protein-binding dyes. If you are suspicious that the crystals you have obtained are salt, a grain or two of the dye can be added to the drop. Salt crystals will not absorb these dyes, but protein crystals will (in theory). After some hours or overnight, the dye should be completely concentrated in the crystals if they are protein. Keep in mind that this test is by no means definitive. The staining is dependent on how well your particular protein binds the dye, the solvent content of the crystal, and the crystallization conditions. At low pH, protein crystals might require days to be stained or they might not get stained at all.

To verify that the crystals are protein and not salt, I find the "snap" method more reliable. Take a needle and try to crush the crystal. Salt crystals are so hard to break that you can often hear them snap. Protein crystals are much easier to smash.

E

Epitaxial Jumps (*Enrico Stura*)

This is a method for stimulating nucleation of a new crystal form on a crystal face. It consists of changing the crystallization conditions by increasing the precipitant concentration or changing the type of precipitant and seeding. See Chapter 14, *Seeding*.

F

Filtration (*Naomi Chayen*)

An alternative to centrifugation of the protein solution is filtration. This method also requires a table centrifuge, but centrifugation is used to drive the solution through a filter. It is common practice to filter the protein solution through a 0.22 μm filter prior to setting up crystallization trials, to remove dust, microbes, aggregated protein, etc. Filters with different molecular weight cut-offs (so-called ultrafiltration membranes) can also be used depending on how rigidly one wants to clean the protein solution.

For screening experiments:

Don't filter at all because this may remove a heterogeneous nucleant crucial for the crystallization of your sample.

For optimization of the conditions for crystallization:

First try filtering the protein solution through a 0.22 μm filter. If too many small crystals are produced filter your solution through a 0.1 μm filter, and if you are still not getting single crystals filter the solution through a 300,000 MWCO ultrafiltration membrane. Mix the filtered protein solution with your precipitating agents and dispense your trials.

Alternatively, for further cleaning, mix your protein solution with the crystallizing agents and then filter the mixture through the various filters before setting up the experiments.

If no crystals are produced at all after filtration, repeat the experiment using a slightly higher (1-3%) concentration of the protein or precipitating agent.

Buffers and precipitant reagents for crystallization should always be filtered through 0.22 μm when they are made. This will prevent microbial growth in them during storage.

Filtration (*Madeleine Riès-Kautt*)

Before filtering a protein solution:

- Wash the filter with the same solution (except the protein)
- Filter a small aliquot of protein solution and verify the protein concentration before and after filtration. Some proteins adsorb to certain types of filter membranes (cellulose acetate, nylon, polysulfone, etc.) This is protein-specific and should be checked for a given type of filter.

G

Gap (*Enrico Stura*)

A gap between the cover slip and the well is the most common reason for a vapor diffusion drop drying out or for observing a precipitate where a clear drop is expected.

Gels (*Naomi Chayen*)

Crystallization in gels reduces convection and in many cases larger, better ordered crystals can be grown. Both agarose and silica gels are used for crystallization. As the gel content increases in silica gels, the number of nuclei decreases while in agarose gels the number increases.[1]

Reference:

1. Provost K, Robert M-C: **Application of gel growth to hanging drop technique.** *J Cryst Growth* 1991, **110**:258-264.

Glycerol (*Madeleine Riès-Kautt*)

Glycerol is very efficient in reducing twinning, probably because it increases the solubility of the protein. See *Twinning*.

Because of this solubility increase, some protein should be added to the cryo-solution when testing glycerol as a cryo-protectant. The amount of protein to add should be about twice the residual protein concentration in the supernatant. (See Chapter 10, section 10.7.) If the residual protein concentration cannot be determined, as a rule of thumb, use one-fifth of the initial protein concentration in the drop (the concentration before addition of the reservoir components).

Glycerol; contamination from membranes (*Terese Bergfors*)

Ultrafiltration membranes from almost all manufacturers (Amicon, Spectrapor, etc.) are treated with glycerol and sometimes sodium azide. If concentration or filtration of your protein through these MWCO (molecular weight cut-off) devices is the final step before crystallization trials, you may unintentionally be including glycerol as part of your setup. Read the manufacturer's operating instructions for how to rinse the membranes.

Glycerol; how to pipette (*Elspeth Garman*)

To pipette glycerol accurately for making up cryo-solutions, warm it by placing the tube in a beaker of hot water first. This lowers the viscosity and makes it much easier to pipette.

Glycerol; how to pipette (*Madeleine Riès-Kautt*)

Alternatively, you may work with an 80% v/v stock solution, which is much less viscous.

H

Habit (*Enrico Stura*)

Habit is the characteristic crystalline form of a protein or mineral. A change in habit is not necessarily related to a change in space group, but confirm by X-ray analysis or with a series of seeding experiments. See section *Epitaxial jumps* in Chapter 14, *Seeding*.

His tags; can you crystallize proteins with them (*Terese Bergfors*)

For a discussion of the pros and cons of removing His tags before crystallization, see Hampton Research's website under "info and ideas." See also *Tags and tails*.

I

Imidazole malate (*Enrico Stura*)

Imidazole malate is a particularly useful buffer because it covers a pH range from 4.5-9.0 and does not crystallize easily. Note that it chelates metals. Make up a 2 M imidazole solution and titrate it with 2 M malic (not maleic) acid.

Reference:

1. Stura EA, Nemerow GR, Wilson IA: **Strategies in the crystallization of glycoproteins and protein complexes.** *J Cryst Growth* 1992, **122**:273-285.

Ionic strength (*Madeleine Riès-Kautt*)

Ionic strength $I = 1/2 \; (\Sigma \; C_i \times Z_i^2)$, where C_i is the molar ion concentration and Z_i the valence of the ion.

For a 1:1 electrolyte, like NaCl, $I = C$.

For a 2:1 electrolyte, like ammonium sulfate, $I = 3 \; C$.

This means that a 1 M ammonium sulfate solution has the same ionic strength as a 3 M NaCl solution.

J

Jeffamine (*Lesley Lloyd Haire*)

The Jeffamine polyoxyalkyleneamines contain primary amino groups attached to the terminus of a polyether backbone. They are thus "polyether amines." The polyether backbone is based on either propylene oxide (PO), ethylene oxide (EO), or mixed EO/PO. The Jeffamine family consists of monoamines, diamines, and triamines which are available in a variety of molecular weights, ranging from 89 to 6,000. The Jeffamine monoamines are designated as the M-series. The M is representative of the fact that they are **M**onoamines. The number designation after the letter M represents the approximate molecular weight. The D-series are diamines and the ED-series are aliphatic diamines structurally derived from propylene oxide-capped polyethylene glycol. These compounds are produced by the Huntsman Corporation (formerly the Texaco Chemical Company) for industrial applications such as lubricants and fuel additives. They were first used as crystallization agents by Duilio Cascio (University of California Los Angeles). In 1992, their use as crystallization agents was brought to the attention of the crystallization community at large by Alexander McPherson at the 4th ICCBM.[1] The first protein structure solved using Jeffamine was that of xylose isomerase from *Thermoanaerobacterium thermosulfurigenes* and the crystallization was described by Lloyd *et al.*[2] The crystals grown from Jeffamine ED 4000 were found to be of better quality than those grown from PEG 4000. (ED 4000 was a generous gift from Texaco to L. Lloyd.) Crystallization of xylose isomerase from *Thermatoga neapolitana* using the microbatch method and Jeffamine ED 4000 is described in reference.[3]

The following Jeffamine products utilized as crystallization agents are available directly from the Huntsman Corporation (who will supply free samples; check the web at www.huntsman.com), or from Fluka (look under *Jeffamine* in Fluka's catalog). Hampton Research also sells pre-formulated M-600 crystallization solutions and ED-2001 (see their *Optimize* reagents).

ED-series:

ED-600 is liquid at room temperature.
ED-900 partially solidifies at temperatures below 25°C.
ED-2001 is a white waxy solid, melting point 40-45°C.
ED-4000 is also a waxy solid. This compound is not commercially available.

D-series:

D-230 and D-400. Both are liquid at room temperature.

M-series:

M-600. This compound is used in the Hampton Crystal Screen II formulation.

Recipe for 40% w/v Jeffamine ED

1. The ED series can be a solid wax and must be melted first. Warm to 37°C.
2. Pour out 40 ml and add 40 ml distilled water.
3. Add 10 ml HCl—the solution becomes hot and the pH is still around 9.
4. Cool to room temperature.
5. Adjust the pH very carefully to 7.0. The pH changes very rapidly as you approach neutrality.
6. Make up to a final volume of 100 ml to give a 40% solution.
7. Store in the dark at 4°C.

N.B. Check the solution pH if using *any* of the Jeffamine compounds listed above because all these compounds are alkaline.

References:

1. McPherson A: **Two approaches to the rapid screening of crystallization conditions.** J *Cryst Growth* 1992, **122**:161-167.
2. Lloyd L, Gallay O, Akins J, Zeikus J: **Crystallization and preliminary X-ray diffraction studies of xylose isomerase from** *Thermoanaerobacterium thermosulfurigenes* **strain 4B.** *J Mol Biol* 1994, **240**:504-506.

3. Chayen NE, Conti E, Vielle C, Zeikus JG: **Crystallization and initial X-ray analysis of xylose isomerase from** *Thermatoga neapolitana*. *Acta Cryst* 1997, **D53**:229-230.

K

Kinetics of drop/reservoir equilibration (*Madeleine Riès-Kautt*)

The equilibration of the drop against the reservoir can be affected in many ways:

- The larger the drop, the slower the equilibration.
- The greater the initial difference in salt concentration between the drop and the reservoir, the faster the equilibration. In practice this means that in a row of drops over reservoirs ranging from 0.5 to 2.0 M salt, the initial salt concentration ranges from 0.25 to 1.0 M. The equilibration rate will be slowest over the 0.5 M reservoir and fastest over the 2.0 M.
- When working with PEG in vapor diffusion drops, the equilibration is *very* slow (it can take weeks to equilibrate). This has two practical consequences:
  1. If crystals appear within a few days, the technique is more like the batch method than vapor diffusion as equilibration has surely not been achieved;
  2. To increase the equilibration kinetics, some salt gradient between drop and reservoir should be introduced.

Reference:
1. Luft JR, DeTitta GT: **Kinetic aspects of macromolecular crystallization.** *Methods Enzymol* 1997, **276**:110-131.

Kinetics of drop/reservoir equilibration (*Naomi Chayen*)

The distance between the drop and the reservoir is also a major parameter affecting the equilibration rate. The longer the distance, the slower the equilibration. Special plates (Z/3 plates) have been made for varying this distance.[1] In Linbro/Cryschem type plates it is not possible to manipulate the drop to reservoir distance to any significant extent. Instead, an oil barrier can be used to slow down the equilibration. See Chapter 16, *Oils for Crystals,* p. 173.

Reference:

1. Luft JR, Arakali SV, Kirisits MJ, Kalenik J, Wawrzak I, Cody V, Pangborn WA, DeTitta GT: **A macromolecular crystallization procedure employing diffusion cells.** *J Appl Cryst* 1994, **27**:443-452.

L

Lanthanides *(Enrico Stura)*

Lanthanides are useful as heavy atom derivatives and as additives in the crystallization of proteins that bind divalent metals.

Ligands, removal of excess *(Terese Bergfors)*

Plate 27 shows crystals of CRABP I complexed with its native ligand, all-trans retinoic acid.[1] When exposed to light, all-trans retinoic acid isomerizes to the 13-cis form. Crystals exposed to light suffered rapid deterioration in diffraction quality (1.9 to 6Å), so all crystallization trials and subsequent data collection were performed in the dark. The stereomicroscope was fitted with a special filter to permit mounting and examination of the light-sensitive crystals.

The protein was purified in its apo form and incubated overnight in the presence of all-trans retinoic acid. Crystallization trials with the liganded form of the protein produced only poor quality crystals unless the complex was further purified on a G50 (size exclusion) column to remove excess ligand and ethanol (in which the retinoic acid was dissolved).

Although this is an example where it was necessary to remove excess ligand, one can usually begin by simply mixing the protein with its ligand in a 1:1 molar ratio just before setting up the drop. This ratio can be manipulated as one of the parameters, e.g., try a 1:5 or 1:10 protein to ligand molar ratio.

Reference:

1. Bergfors T, Kleywegt GJ, Jones TA: **Crystallization and preliminary X-ray analysis of recombinant bovine cellular retinoic acid-binding protein.** *Acta Cryst* 1994, **D50**:370-374.

M

Microgravity (Space) *(Naomi Chayen)*

Microgravity eliminates sedimentation and convective mixing, thus offering a more homogeneous growth medium compared to growth conditions on earth. It is therefore likely to improve the degree of perfection of the crystals.

The reported rate of success for crystallization in microgravity is about 20%, i.e., 20% of the crystals grown in Space show improvement (in size or diffraction quality) compared to crystals grown under similar conditions on earth. Hence, the value of crystallization under microgravity is still debated. Recent development of new apparatus (led by the European Space Agency), which facilitates monitoring of the crystal growth process in Space, is enhancing the scope and efficiency of experimentation in microgravity.

Flight opportunities are available through NASA, ESA, and the National Space Agencies.

MPD *(Madeleine Riès-Kautt)*

MPD (2-methyl-2,4-pentanediol) is an alcohol often used in protein crystallization. It is important to realize that such alcohols are azeotropic in water (the concentration of the two substances in the vapor state is different from the liquid mixture except at the azeotropic point). This is the reason it can cause peculiar behavior in vapor diffusion drops, such as the drops swelling instead of shrinking.

MPD *(Terese Bergfors)*

MPD degrades and should be stored in the dark at 4°C because the purity can influence the crystal quality. Measure the absorbance at

280 nm: high quality MPD has an OD_{280} ≤0.01. If the optical density is higher, buy new MPD or purify it. Purification can be carried out by fractional distillation or adding 20% w/w charcoal overnight, followed by filtration.[1]

Reference:

1. Zeppezauer M: **Formation of large crystals.** *Meth Enzymol* 1971 **XXII**:253-269.

Multi-drops *(Terese Bergfors)*

Up to 4 sitting or hanging drops of 10 µl each can be placed on an 18 mm cover slip and equilibrated against the reservoir. This is a useful technique for parallel screening of additives, mutants, buffers, etc. Be sure to include a reference mark on the cover slip to know which drop is which.

Reference

1. Bruns CM, Karplus PA: **The multi-drop approach: more efficient screening of crystallization conditions.** *J Appl Cryst* 1995, **28**:242-243.

Multiple crystal forms

see *Polymorphs*

N

Nucleants *(Naomi Chayen)*

Once your crystallization solution (containing the protein to be crystallized and the crystallizing agents) has been filtered, you can add external nucleants to your trial in a controlled manner.

Any external substance added to your crystallization trial can act as a nucleant. This could be a protein crystal seed, an inorganic crystalline surface, latex beads, or even just dust. Use your imagination and try anything—it may prove to become a "universal nucleant."

P

PEG; how to make a 50% (w/v) PEG solution (*Terese Bergfors*)

A 50% w/v PEG solution will have a different concentration than a 50% w/w solution. (See *Units of concentration.*). Whatever your preference for making PEG, be sure to specify if it is %w/v, %w/w, or %v/v to avoid confusion. To make a 50% w/v PEG solution, dissolve 50 g of PEG in water and bring the *final volume* to 100 ml. A common mistake is to *add* 50 g PEG *to* 100 ml: The final volume will be >100 ml. To make w/w or v/v solutions, see below.

PEG; how to make a 60% (w/w) PEG 2000 solution (*Enrico Stura*)

High molecular weight (1000-20,000) PEG solutions are best made as w/w. Weigh 60 grams of PEG 2000 and add 40 grams (or ml) water, (include 0.1% azide). Microwave in a beaker; stop as soon as it starts to boil. Mix well while hot and store when cooled down.

PEG; how to make a 60% (v/v) PEG 600 solution (*Enrico Stura*)

Liquid PEGs (molecular weight <400) are best mixed as v/v. PEG 600 is a wax at room temperature. Microwave PEG 600 (or use a hot water bath) to liquefy it. Mix 60 ml of PEG 600 with 40 ml water while warm. Mix well.

PEG; store in the dark (*Terese Bergfors*)

PEG (polyethylene glycol) is commercially produced in large quantities and the quality varies enormously from brand to brand and even batch to batch. Always buy the best quality PEG (gas chromatography quality when available) and make a note of the manufacturer/batch. Mixing brands in your crystallization trials will practically guarantee irreproducibility of your results.

The PEG should be stored in a dark bottle. Some people even flush the PEG solution with nitrogen and store aliquots at -20°C. Light and oxygen accelerate the decomposition of the PEG, which often contains peroxides and aldehydes or aldehyde precursors which hydrolyze to aldehydes. The acetaldehyde has a characteristic odor which is detectable upon opening the bottle.

References:

1. Ray WJ, Puvathingal JM: **A simple procedure for removing contaminating aldehydes and peroxides from aqueous solutions of polyethylene glycols and of nonionic detergents that are based on the polyoxyethylene linkage.** *Analyt Biochem* 1985, **146**:397-312.
   Explains the types of contaminants in PEG; gives procedures for purifying PEG.

2. Jurnak F: **Induction of elongation factor Tu-GDP crystal polymorphism by polyethylene glycol contaminants.** *J Mol Biol* 1985, **185**:215-217.
   This paper shows how different batches of PEG 3350 gave different crystal forms.

pH (*Madeleine Riès-Kautt*)

The pH of a protein solution, even of a drop ≥5µl, can be measured with a micro-electrode (Radiometer MIXC 410). For measuring the pH of larger volumes (= 20 µl, e.g., for reservoir solutions) a less fragile micro-electrode (Mettler Toledo InLab 423) can be used. (See also Appendix A1, *Good-to-Have Gizmos*.)

pH (*Naomi Chayen*)

I have found that if one places a standard pH electrode in a reservoir solution containing, for example, 1.5 M ammonium sulfate, it can take over 10 minutes for the electrode to recover, even after thorough washing. To avoid this, I take a small aliquot (around 50 µl) from the reservoir, dilute it ten-fold, and measure the pH. The pH is unaffected by the dilution and the electrode does not get a shock.

Phase separation (*Naomi Chayen*)

A common occurrence in crystallization trials is the formation of phase separation, which looks like oily droplets inside the crystallization drop. This often happens when detergents are present or when PEG and salt are used together. Do not panic if this happens; crystals can still grow.

Phase separation (*Terese Bergfors*)

The large crystal, Plate 28, of platelet derived growth factor grew in a vapor diffusion hanging drop about 1 week after phase separation occurred. The drop contained 1 M LiCl, 0.4% ß-octyl glucoside and 10 mg/ml protein in MES, pH 6.0. Phase separation can present problems when mounting the crystal (see Chapter 13, section 13.2). Phase separation can sometimes be avoided by lowering the concentrations of protein and precipitant. See the phase diagram in Chapter 10, Figure 10.1 .

Polymorphs (multiple crystal forms) *(Gerard Kleywegt)*

Previously, polymorphs were often considered a nuisance, but nowadays people realize that having multiple crystal forms can be extremely useful later on in the structure determination (e.g., for density averaging to improve phases.) Also, sometimes parts of a structure that are disordered in one crystal form may have fantastic, rock-solid density in another. Collect and keep datasets of all crystal forms that diffract to 3.5Å or better.

Reference
1. Kleywegt GJ, Read R: **Not your average density.** *Structure* 1997, **5**:1557-1569.

R

Recycle; how to recycle your protein or crystals from drops
(Terese Bergfors)

It is quite possible to recycle crystals from crystallization trials. Re-dissolving and re-crystallizing small crystals can act as an extra purification step which sometimes produces large, useful crystals. Simply adding water to the vapor diffusion drop may make the crystals dissolve; upon re-equilibration against the reservoir, better crystals may result.

Sometimes it is quite difficult to get the crystals to dissolve, especially the older they are, since the surfaces tend to become cross-linked. In these cases, or if you do not have any crystals at all but wish to recycle the protein from your drops, you can add water, buffer, or mother liquor to the drops to increase the volume to a reasonable working amount. Pool the drops in a microcentrifuge tube and if there are any crystals, crush them. Centrifuge the solution 1-2 minutes in a micro-centrifuge at 13,000 g to pellet undissolved precipitate. Check the protein concentration in the supernatant and exchange the buffer now to remove the precipitants and additives that were in the pooled drops. (See Appendix A1, *Good-to-Have Gizmos*: how to exchange buffers in small volumes.) PEG is very difficult to get rid of by dialysis; MicroSpin columns are not sufficient for its removal either. Run the protein on a Superose 12 (Pharmacia) or similar size exclusion column to get rid of PEG $\geq$ 2000.

References:

1. 1. McRee DE: *Practical protein crystallography.* New York: Academic Press, Inc.; 1993:5-6.

2. Ducruix A, Giegé R: *Crystallization of nucleic acids and proteins. A practical approach.* Oxford: IRL Press; 1992:86.

These two works present other protocols for ultrapurification of small crystals.

Reducing agents (anti-oxidants); protection of free cysteines
(*Madeleine Riès-Kautt*)

Free cysteines are likely to form intra- or intermolecular disulfide bridges (especially at pH >6) leading to heterogeneity or aggregation. This can be avoided by reducing the dissolved oxygen in the solutions or by adding anti-oxidants.

Oxygen can be replaced by nitrogen (or helium, etc.) by degassing the solutions in a desiccator under vacuum for 10-20 minutes. At the end, introduce nitrogen into the desiccator to bring it to atmospheric pressure.

The efficiency of anti-oxidants is variable and decreases with time.

- β-mercaptoethanol (β-me) has one SH group which interacts directly with the SH group of the free cysteine. Its efficiency decreases the fastest (2-3 days) of the three anti-oxidants listed here. It is liquid, and the odor is strong; work under a hood. When working with vapor diffusion (hanging drops), it is best to refresh the anti-oxidant by adding it to the reservoir at least once a week.

- Dithiothreitol (DTT) bears 2 SH groups. It is more efficient than β-me and it lasts for about 3-7 days (especially at 4°C). It cannot be renewed in hanging drops, but is most often used for crystallization by dialysis where the reservoirs can easily be changed to refresh the anti-oxidant.

- Tris(2-Carboxyethyl)-Phosphine Hydrochloride (TCEP-HCl) is much stronger than β-me or DTT (it cleaves disulfide bridges if there are any in the protein), and lasts for 2-3 weeks.

  Warning: It acidifies the solution, which must be adjusted to the right pH after preparing the solutions.

The anti-oxidants are usually used at 1 mM during the preparation of the protein. For crystallization, check the protein concentration and the number of free cysteines (for example, a 20 mg/ml protein of 20 KDa with 6 free Cys is 1 mM and requires at least 6 mM of anti-oxidant).

Reducing agents (*Elisabeth Sauer-Eriksson*)

Reducing agents are powerful crystallization additives and I recommend that you test them in your first crystallization trials. This can be done by placing two drops on each cover slip and adding DTT to one of them. β-me is volatile so it cannot be used in this multi-drop method since it will diffuse through the vapor phase into the other drop. To test β-me, add 1 μl to the 1 ml reservoir solutions in the 24-well plates. β-me and DTT are quite different so both compounds should be tested. There are many examples where the choice of reducing agents turned out to be crucial in the crystallization.

β-me can be added to proteins without cysteines. In T4 lysozyme, the two Cys residues were mutated to Ser. Nonetheless, β-me was required for successful crystallization. The β-me bound to the OH-group of one of the Ser molecules and the SH-group of the β-me formed a disulfide bond with a symmetry related copy of itself over a crystallographic two-fold axis. Substituting reduced β-me for oxidized β-me reduced the crystallization time of the T4 lysozyme mutant from 6 months to 1-2 weeks.[1] There are also cases where β-me has been an essential additive to the crystallization of proteins without any cysteines although no β-me molecule was identified in the resulting structures.

To prevent inactivation of your reducing agents by trace metal compounds, if possible, include some EDTA in your trials. (Do not use EDTA if your protein requires metals for its activity.) A typical combination is 1 mM DTT and 1 mM EDTA. At pH 8-9 the half life of DTT is only minutes, so β-me is a better choice when working at alkaline pH.

β-me is also used instead of DTT when purifying His-tagged proteins on Ni-agarose columns, because the DTT reduces the nickel ions.

L-cysteine is sometimes recommended as a reducing agent additive. I have found its usefulness limited because it easily crystallizes as small hexagonal plates.

Reference:

1. Eriksson AE, Baase WA, Matthews BW: **Similar hydrophobic replacements of Leu99 and Phe153 within the core of T4 lysozyme have different structural and thermodynamical consequences.** *J Mol Biol* 1993, **229**:747-769.

Refractive index (*Madeleine Riès-Kautt*)

Sometimes crystallization conditions are not reproducible because slight changes occur during the preparation of the stock solution of the crystallization agents. An easy control is to measure and compare the refractive indices of the solutions with a refractometer. Note the temperature, because the refractive index is very sensitive to temperature.

In addition, it is easy to verify the refractive index (with 20-40 µl) of the reservoir solution. Sometimes the cover slip has not been thoroughly sealed and evaporation has occurred, especially at high temperatures.

Robots; for crystallization (*Naomi Chayen*)

A number of papers describing various robots, mostly for setting up vapor diffusion trials, have been published and several laboratories have built their own automated dispensers which are used in-house. However, currently there are only two automated systems commercially available.

One system, the C-200 Robotic Workstation made by Cyberlab, can perform vapor diffusion experiments. The other, Oryx, made by Douglas Instruments, can perform both microbatch and vapor diffusion.

Both systems can be used for screening and optimization experiments. They each offer a mode for screening to find initial conditions using either pre-mixed, ready-made solutions (such as the Hampton Research kits) or they can mix custom-tailored screens designed by the experimenters using the systems' software. The optimization is performed as matrix screens around the conditions of interest. Protocol documentation and calculation of the amounts of crystallizing agents and protein are software features in both systems. Both robots will adjust the pipetting rate to compensate for viscous precipitants.

Specific features of the Cyberlab C-200 Robotic Workstation:

This is a fully automated workstation with integrated software for setting up hanging drop experiments. (A sitting drop option will soon be incorporated.) The workstation exactly replicates the manual method. It prepares 24-well plates using disposable pipette tips, greases the plates, and places the cover slips over the wells so no user intervention is necessary. Up to six plates can be prepared at a time. The workstation handles volumes down to 2 μl.

Specific features of the Oryx Automated System:

Oryx is an automated system for setting up sitting drops, hanging drops, and microbatch experiments. The main feature of Oryx in the mode for dispensing microbatch trials is its ability to accurately dispense final trial volumes of 1 μl. It does so by dispensing the samples under oil (see Chapter 16, Figure 16.1) through a specially designed fluoropolymer microtip. In the case of vapor diffusion trials, Oryx uses disposable tips to dispense the reservoirs and a fine microtip to dispense the reservoir and protein solutions onto the cover slips. Oil (but not grease) for sealing the wells is automatically applied to the Linbro plates and the cover slips are placed on the plate by hand.

Suppliers' details are found in the appendix or see the respective websites:

www.douglas.co.uk
www.cyber-lab.com

S

Silanization (*Terese Bergfors*)

Cover slips for hanging drops should be silanized to make the glass hydrophobic.

Several suppliers sell already-silanized glass cover slips and plastic cover slips (which need no silanization) are available from Nunc. If you silanize the glass yourself, many silanizing solutions are available, e.g., Prosil 28 (Hampton Research) or Repel-Silane ES (Pharmacia cat. no. 17-1332-01). A cost-effective alternative for making your own silane solution is presented here:

Solutions required:

2% dichloromethyl silane (Sigma cat. no. D-3879)
1,1,1-trichloroethane (Sigma cat. no. T-4678)

How to make the silanizing solution:

1. All silanization work must be done in a fume hood. Wear gloves.
2. In a one liter GLASS graduated cylinder, pour 20 ml dichloromethyl silane.
3. Add 980 ml trichloroethane. Store in a dark bottle. The solution can be re-used until it turns yellow.

How to silanize glass cover slips:

1. In a so-called "crystallization dish" or any low, wide glass beaker or dish, place about 100 glass cover slips. They should not overlap each other too much or they will not be properly coated.
2. Fill one-third of the bowl with the silane mixture. The trichloroethane is highly volatile and much will evaporate.
3. Put the bowl on a shaker table under gentle agitation, in the fume hood, for a minimum of 30 minutes. Do not leave it overnight though, as the solution will evaporate by morning leaving a gummed-up mess.
4. Pour off and recover the silanizing solution for re-use. A glass Urbanti funnel is ideal for pouring out the cover slips and solution.
5. Replace the cover slips in the dish and rinse them thoroughly with ethanol in a series of three washes over a period of at least 15 minutes.
6. Dry the cover slips in the lab oven usually used for drying glassware. Drying can also be done in a microwave oven and will be much faster (10-15 minutes). Remember to put a sepa-

rate beaker of water inside the microwave oven so that you do not destroy it (by running it to complete dryness).

How to silanize glass or quartz X-ray capillaries:

Silanizing the capillaries greatly facilitates removal of excess mother liquor around the crystal because the liquid beads up easier.

1. Open a package of X-ray capillaries. Snip off the end tip of each capillary.

2. Place them one by one in a tall, 100 ml glass beaker and cover them completely with silanizing solution. The capillaries will suck up the solution. Let them stand for a minimum of 30 minutes.

3. Rinse them thoroughly in ethanol and dry them as described above. If the capillaries are not thoroughly rinsed, a white smudge accumulates on the inside. This is highly detrimental to the crystal. Discard any such capillaries or rinse them better.

How to silanize plastic:

Plastic is inherently hydrophobic and usually does not require silanization. In cases where the crystals stick to the plastic micro-bridges, it can be worth a try to silanize the plastic. A special aqueous silane is required for plastic.

1. Purchase Aqua-Sil (Pierce cat. no. 42799).

2. Take 1 ml Aqua-Sil and add 99 ml water to give a 0.2% solution (because the fluid in the bottle is 20%). This dilution must be prepared fresh each time.

3. To preserve the remaining, expensive stock solution, gently blow nitrogen gas into the bottle for about 10 minutes, then re-seal.

4. Silanize the plastic microbridges, etc., according to the instructions included with the Aqua-Sil.

5. Rinse the plastic thoroughly in distilled water to remove traces of the Aqua-Sil, and dry in a microwave oven.

Sitting drops (*Terese Bergfors*)

To slow down the rate of vapor diffusion, replace your hanging drop with a sitting drop.

Reference:

1. DeTitta GT, Luft JR: **Rate of water equilibration in vapor-diffusion crystallization: dependence on the residual pressure of air in the vapor space.** *Acta Cryst* 1995, **D51**:786-791.

Spherulites (*Madeleine Riès-Kautt*)

Crushing spherulites with a glass tip can induce nice crystals in the same drop.

Stick; crystals stick to the cover slip (*Torsten Unge*)

It is sometimes unavoidable that crystals stick to the cover slip, regardless of the technique. A simple solution to this problem is to add a few μl of the reservoir solution to the droplet. This small change in ionic strength or precipitant concentration can be enough to elicit the loosening of the crystal. In some cases, the slight increase in temperature when touching the outside of the cover slip with the finger tip is enough to loosen the crystal.

Stick; crystals stick to the cover slip (*Terese Bergfors*)

Never poke the crystal with a needle. Use a cat whisker (see Chapter 14, *Seeding*) to gently pry the crystal loose. Another method is to take waxed weighing paper and cut it into a manageable size. Fold it into a paper airplane. This creates a point to glide underneath the crystal. The paper is waxed and therefore does not absorb mother liquor.

Stir; to stir or not to stir your drops (*Terese Bergfors*)

After making the crystallization drop (protein plus precipitant), the question arises of whether you should stir the drop or not. Stirring will increase the number of nucleation sites. Therefore, stirring can be a good idea when screening for the initial crystallization conditions, but it is not recommended when optimizing. Whichever strategy is chosen (stirring or not stirring), what is probably most important is consistency in approach from experiment to experiment.

To stir or to mix? Stirring is done by using the pipette tip, as you would with a stick or spoon, to blend the ingredients of the drop. Mixing is done by aspirating the drop with the pipette a few times. In either case, avoid introducing bubbles into the drop.

Along these same lines, one should always add the precipitant to the protein and not the other way around. This is to avoid high, local concentrations of precipitant which might precipitate the protein.

T

Tags and tails (*Enrico Stura*)

Tags, His-tags, peptide-tags, and additional tails at the end of the natural protein sequence pose a greater problem the longer the tail is. It is not expected that such unnatural additions will be ordered. Hence they contribute to conformational heterogeneity of the protein. A short tail, not involved in crystal contacts, will be well tolerated. Proteins with long tails tend to give rise to poorly diffracting, and even anisomorphous, crystals.

Start the screening with tagged proteins whatever the tag length. These proteins are generally easier to purify and are available in larger amounts. Use the resultant crystals to seed experiments set up with the tagless protein, which is probably less available and less pure. Seeding will increase the number of conditions under which crystals are obtained.

Removing a long tag or deleting a disordered part of a natural protein will often result in better diffracting crystals. Ligands binding can also order otherwise disordered parts of a protein. A good example of this is glycinamide ribonucleotide transformylase crystals for which it improved the diffraction limit from 3.0Å[1] to 1.96 Å.[2]

References:

1. Chen P, Schulze-Gahmen U, Stura EA, Inglese J, Johnson DL, Marolewski A, Benkovic SJ, Wilson IA: **Crystal structure of glyci-**

namide ribonucleotide transformylase from Escherichia coli at 3.0 Å resolution. A target enzyme for chemotherapy. *J Mol Biol* 1992, **227**:283-292.

2. Klein C, Chen P, Arevalo JH, Stura EA, Marolewski A., Warren MS, Benkovic SJ, Wilson IA: **Towards structure-based drug design: crystal structure of a multisubstrate adduct complex of glycinamide ribonucleotide transformylase at 1.96 Å resolution.** *J Mol Biol* 1995, **249**:153-175.

Tungstate (*Enrico Stura*)

Tungstate is a phosphate mimic. It is normally used as an additive for proteins that bind phosphate, in particular phosphatases and kinases. It is also useful as a heavy atom for phasing. Higher concentrations than vanadate are often needed to achieve the same effect.

Twinning (*Naomi Chayen*)

Twinning is when two or more crystals grow together, either stuck to one another or inside one another.

Dioxane (0.5 to 2%) is a common agent for reducing twinning. A word of caution: Dioxane is a small volatile organic molecule and therefore cannot be used with the oil-based microbatch method.

Glycerol has also been mentioned here in *A-Z Tips* as an anti-twinning additive. Crystallization in gels may also help.

U

Units of concentration (*Madeleine Riès-Kautt*)

Very different units are used to indicate the concentration of a solution:

Molarity (number of moles per liter of solution) is probably the most accurate unit to use.

% w/v and % v/v are similar to molarity as long as it is a weight (or volume) of solute *in* a given volume and not a weight (or volume) *plus* the solvent volume. This is important when preparing a concentrated protein solution (mg/ml): A solution prepared with 50 mg lyophilized protein powder *plus* 1 ml or *in* 1 ml are very different. It also affects the concentration of dilutions obtained from a stock solution: diluting a 2% (weight or volume) stock solution by ten will give a 0.2% solution only if the stock solution was 2% (weight or volume) *in* the solvent, and not otherwise.

% saturation. This unit is still often used for ammonium sulfate. The concentration of a saturated stock solution depends on the temperature. Depending whether it is kept at room temperature or in the cold room, the concentration will be very different.

V

Vanadate (*Enrico Stura*)

Vanadate is a phosphate mimic. It is a useful additive for proteins that bind phosphate, in particular phosphatases and kinases.

Vibration (*Naomi Chayen*)

Vibration can cause excess nucleation, leading to the formation of large numbers of low quality crystals. My best crystals often grow while I am away from the laboratory for a week or two. When optimizing, restrain your curiosity. Do not look at the crystals every couple of days—they need to be left in peace to grow. On the other hand, when screening, or if you have no crystals, move the trays around. The vibration may trigger some nucleation.

Vibration (*Terese Bergfors*)

My best crystals grow during the Christmas holidays, when all the *other* people are away from the laboratory. Opening and shutting the door to the crystallization room can be a significant source of vibrations. Add a "door closer" to the door. This is a hydraulic arm, available at hardware stores, that regulates how smoothly the door closes.

W

Water (*Terese Bergfors*)

The source of water for the crystallization solutions is critical. Many apocryphal (but true) stories exist about how crystals grew in one place but could not be repeated in a different laboratory unless the same water was used. Different water purification systems do not produce identical water. For example, glass-distilled water has a pH of 5.5. One laboratory may be purifying the water by double distillation; another may be doing it by deionization or reverse osmosis. This is one of the factors contributing to the irreproducibility of results with water from different laboratories.

Another important consideration is the storage of highly purified water. Extremely pure water leaches substances from the glass or plastic in which it is stored. A study cited by Millipore[1] showed that after several weeks' storage in plastic, the total organic carbon content of the water was at nearly the same level as in ordinary tap water. Table T.1 shows the effects of storing the water in glass.

Table T.1. Contamination levels after storage in Pyrex

	in redistilled water (ppb)	after 2 weeks in Pyrex (ppb)
Zn	14	46
Pb	9	30
Cu	5	12
Fe	9	45
Al	10	102

Reprinted by permission from Millipore.

Moral:

When preparing buffers and precipitants for crystallization, always use freshly purified water, not water stored in a carboy or distilling vessel for days/weeks.

Reference:

1. Millipore Technical Brief Cat. No. EU412/U. March, 1994. **"Ultrapure ion-free/organic-free water for trace analysis"**.

X

X-axis (*Enrico Stura*)

The precipitant concentration is generally plotted along the X-axis of the solubility curve (phase diagram). Screening for crystallization conditions far along the X-axis at high precipitant concentration will favor the growth of crystals with minimal solvent content. This, in turn, means tight packing, well-ordered crystals, and high-resolution data.

Y

Y-axis (*Enrico Stura*)

The protein concentration is generally plotted along the Y-axis of the solubility curve. In the initial screening it is important to maximize the amount of protein in the supersaturation state, i.e., work at a high protein concentration. Crystals obtained in the initial screening can be used to seed into lower protein concentration conditions during optimization.

Z

Zinc (*Enrico Stura*)

Zinc (5 mM) is a useful additive for the crystallization of proteins and is included in many screens. It reduces the solubility of most proteins. It is unfortunate that many zinc salts are poorly soluble and conditions containing this ion often yield salt crystals.

Appendices

Good-to-Have Gizmos

Terese Bergfors

Uppsala University, Uppsala, Sweden

Here are some recommendations for useful gizmos. (Suppliers' addresses can be found in Appendix A3.)

A1.1. Temperature Logger for the Crystallization Room

The perfect crystallization room is vibration-free with constant temperature and humidity. However, if your crystallization room suffers fluctuations in temperature, a temperature logger (thermograph) will record them. (You may not be able to repeat the variations, but at least you will know what they are.) Omega Technologies Ltd. makes a small logger (4×3×2 cm) at a reasonable price. The temperature is recorded at however frequent time intervals you choose and graphed by downloading into a computer. You might want to buy two and put them in different locations in the room.

A minimum requirement for the crystallization room is a min/max thermometer, which can be purchased at any hardware store. These display the minimum and maximum temperatures in a 24 hour period.

A1.2. Positive Displacement Pipettes

Stock solutions of the high molecular weight PEGs (40-60%) are too viscous to dispense accurately with an ordinary air displacement pipette. It is necessary to use a so-called "positive displacement" pipette which pushes the liquid out by means of a piston, e.g., Gilson's Microman series. For making series of PEG dilutions in the reservoirs of a 24-well tray, there are "repeating" positive displacement pipettes. Two such models which cover volumes from 10 µl to 5000 µl are made by Socorrex and Eppendorf. The Socorrex Stepper 411 is the more ergonomic of the two models since the dispensing is done with a full hand grip rather than with the thumb. The Eppendorf Multipette dispenser 4780, on the other hand, tends not to splash as much when dispensing into small reservoirs. The tips are disposable for these models to prevent cross-contamination.

For repeated dispensing of small aliquots, e.g., the protein solution, I recommend a 50 or 100 µl Hamilton syringe and PB600-1 adaptor. This will enable you to accurately repeat 1 or 2 µl drops, respectively. It is possible to make as small as 0.2 µl drops by fitting the adaptor to a 10 µl syringe.

The other alternative of course is to buy a crystallization robot which will do this tricky pipetting for you.

A1.3. Microelectrode for Measuring pH in the Drop or Reservoir

Until recently it was impossible to find any pH electrodes designed specifically for measuring pH in small volumes. Now there are several electrodes (available from World Precision, Radiometer, etc.) which can measure pH in drops of only a few µl. In many cases, unexpected results can be explained after the pH of the drop is verified. Remember that pH electrodes can be a source of contamination. They release ions (K^+ and Cl^-) into the solution being measured and proteins adhere to the glass. If possible, remove a few µl from the drop for the pH measurement.

A.1.4. Buffer Exchange for Small Volumes (<250 μl) of Protein

There are many reasons why you might want to exchange the buffer of your protein solution---to put it in a better buffer for crystallization trials, to remove salt, or to test pH effects with dynamic light scattering, to name but a few. This can be a problem because of the losses involved in handling such small volumes. Here are five suggestions:

a. "Slide-A-Lyzer®" dialysis cassettes

These are available from Pierce, Inc., with molecular weight cut-offs (MWCO) of 3500, 7000, and 10,000 Daltons, ready to use (the dialysis membrane is pre-washed and free of heavy metal contamination). It is easy to see the protein solution in these cassettes (which look like window frames or holders for picture slides); this can be an advantage if your protein begins to precipitate. There are three models to cover volumes from 100-500 μl, 500 μl-3 ml, and 3-15 ml.

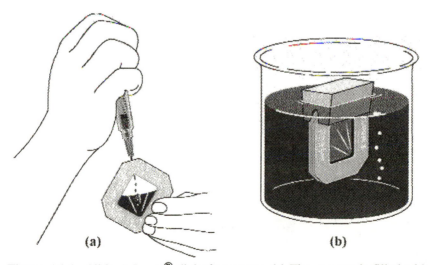

(a)   (b)

Figure A1.1. Slide-A-Lyzer® dialysis cassette. (a) The cassette is filled with the solution to be dialyzed. (b) Cassette with floating device in the dialysis beaker.
Courtesy of Pierce Chemical Company, Rockford, IL, USA.

b. "Tube-O-DIALYZER™"

The tube-o-DIALYZER™ (sold by Chemicon) is available for volumes of 10-250 μl and 200 μl-3 ml. The dialysis membranes are built into the caps of the tubes and are available with MWCOs of 1K, 7.5K, 15K, and 50K. Handling losses are totally eliminated because the protein can be dialyzed and stored in the same tube. Simply replace the dialysis cap with an ordinary microcentrifuge tube lid.

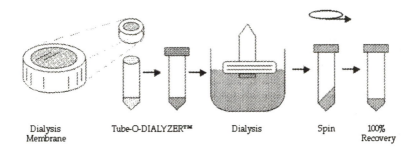

Dialysis   Tube-O-DIALYZER™   Dialysis   Spin   100%
Membrane   Recovery

Figure A1.2. Tube-o-DIALYZER™ microdialysis tubes.
Courtesy of Chemicon International, Inc. Tube-o-DIALYZER™ is a trademark of Geno Technology, Inc. USA.

c. Dialysis tubing fitted over a microcentrifuge tube with an O-ring

By far the cheapest alternative is to make your own microdialysis unit with an O-ring, a microcentrifuge tube, and a piece of dialysis tubing. This is described on Johan Zeelen's website:
    http://www.biophys.mpg.de/kuehlbrandt/zeelen/xtal.html.

d. Dilution and re-concentration in centrifugal microconcentrator

It is also possible to exchange buffer by means other than dialysis. The protein can be concentrated in a centrifugal microconcentrator (e.g., Microcons from Millipore; Nanoseps from Pall Filtron, etc.), re-diluted with the new buffer, then concentrated again, in several cycles to ensure full exchange of the buffer. Follow the instructions provided by the manufacturer.

e. Micro-chromatography columns

These are disposable, 1 ml cartridges of size-exclusion chromatography medium (Pharmacia's MicroSpin, cat. no. 27-5325 or Bio-Rad's Micro Bio-Spin, cat. no. 732-6221). Typically they are used for cleanup and purification of nucleic acids but they also work well for buffer exchange or desalting of most soluble proteins. The protein sample is applied (20-75 µl) and the "column" centrifuged 4 minutes at 1,000xg. The column sits inside a microcentrifuge tube and the protein drains into this.

A1.5. Vortex Mixer Adapted for 24-Well Tissue Culture Plates

If you add the different crystallization reagents (precipitant, buffer, additives, etc.) directly into the reservoirs it quickly becomes tedious to mix them by hand-pipetting. Some vortex mixers are equipped with head attachments that accept 24-well plates. Vortex-Genie 2 (Scientific Industries, Inc.) is such a model. If the vortex is set at the lowest speed, the plate can be shaken without cross contamination between reservoirs. A few minutes is sufficient to blend the different components, even high concentrations of PEG.

A1.6. Automated Grease (Fluid) Dispenser

Applying vacuum grease to the crystallization plates by hand is messy and erratic, making it difficult to insure a complete seal on every well. The Ultra System™ (EFD International) is a compact, air-powered dispensing unit ideal for applying viscous pastes and non-pourable fluids like vacuum grease. This dispenser eliminates waste of grease in addition to making sealing more accurate. The output air pressure is regulated with a foot pedal to provide a uniform flow of grease from the syringe. A barrel loader packs the grease into the syringes. There are four different dispensing tip types. Order the amber (15 gauge) general purpose tip.

Appendix A2

Supplies for Crystallization and Suggested Sources

Books	Hampton Research	Molecular Dimensions	Polycrystal		
Chemicals for crystallization	Hampton Research	Molecular Dimensions	Fluka	Merck	
Cover slips, plastic	Nunc				
Cover slips, silanized	Hampton Research	Molecular Dimensions			
Cryoloops	Hampton Research	Molecular Dimensions			
Detergents	Hampton Research	Calbiochem	Molecular Dimensions	Boehringer Mannheim	Anatrace
Dialysis buttons	Hampton Research	Cambridge Repitition	Sialomed Inc.		
Dialysis, micro quantities	Pierce	Chemicon	Sialomed Inc.		
Dynamic light scattering instruments (classic)	Wyatt				

Continued on next page

Dynamic light scattering instruments	Protein Solutions	Brookhaven Instruments			
Filters, 0.1 and 0.02 µm	Molecular Dimensions	Protein Solutions	Whatman		
Filters, 0.22 µm	Molecular Dimensions	Millipore	Pall Filtron		
Fluid dispensers, automatic	EFD				
Grease for cover clips	Hampton Research	Molecular Dimensions			
Heavy metals, hard to find	Molecular Dimensions	Anatrace	Pfaltz & Bauer	use http://chemfinder.camsoft.com	
Jeffamine	Fluka	Hampton Research	Molecular Dimensions	Huntsman Corp.	
Kits for crystallization	Hampton Research	Emerald BioStructures	Molecular Dimensions		
Micro-tools for mounting crystals	Hampton Research	Molecular Dimensions	Molecular Dimensions		
Oils	Hampton Research	Molecular Dimensions	BDR	ABCR	Merck
pH electrodes, micro	World Precision Instruments	Radiometer	Mettler		
Plates for microbatch	Hampton Research	Molecular Dimensions	Nunc	Douglas Instruments	
Plates for vapor diffusion	Hampton Research	Molecular Dimensions	ICN Biomedicals	Falcon	Costar

Continued on next page

Positive displacement pipettes	Gilson		
Positive displacement pipettes, repeating	Socorrex	Eppendorf	Hamilton
Robots	Douglas Instruments	CyberLab	
Sealing tape, clear	Hampton Research	Molecular Dimensions	
Silane	Hampton Research	Pierce	Pharmacia
Sitting drops, glass rods	Hampton Research		
Sitting drops, microbridges	D.R.O.P.	Hampton Research	
Temperature logger	Omega		
Ultrafiltration concentrators (300 kD, etc.)	Molecular Dimensions	Millipore	Pall Filtron
X-ray capillaries; quartz, glass	Glas Technik & Konstruktion		

Appendix A3

Suppliers' Addresses

Company	Address	Telephone	Fax	E-mail, Website
ABCR GmbH & Co.	P O Box 21 01 35, Hanastr. 29C, D-76151 Karlsruhe, Germany	+49 (0) 721 950 610	+49 (0) 721 950 6180	www.abcr.de
Anatrace, Inc.	434 West Dussel Drive, Maumee, OH 43537, USA	1 (419) 891-3030 Toll free US: 1-800-252-1280	1 (419) 891-3037	anatrac434@aol.com www.anatrace.com
BDH Laboratory Supplies	Poole, Dorset BH15 1TD, UK	+44 1202 660444	+44 1202 666856	export@bdh.com www.bdh.com
Bio-Rad Laboratories	2000 Alfred Nobel Dr., Hercules, CA 94547, USA	1 (510) 741-1000	1 (510) 741-5800	
Boehringer Mannheim GmbH	P O Box 310120, D-6800 Mannheim 31, Germany	+49 621 7590	+49 621 759 2890	
Brookhaven Instruments	750 Blue Point Rd., Holtsville, NY 11742, USA	1 (516) 758-3200	1 (516) 758-3255	wwwsales@bic.com www.bic.com

Continued on next page

Company	Address	Telephone	Fax	E-mail, Website
Calbiochem-Novabiochem Inc.	P O Box 12087, La Jolla, CA 92039-2087, USA	1 (619) 450-9600	1 (800) 776-0999	technical@calbiochem.com www.calbiochem.com
Cambridge Repetition Engineers LTD.	Greens Road, Cambridge CB4 3EQ, UK	+44 223 64655	+44 223 467328	
Chemicon International, Inc.	28835 Single Oak Dr., Temecula, CA 92590, USA	1 (909) 676-8080 toll free US: 1-800-437-7500	1 (909) 676-9209 toll free US: 1-800-437-7502	custserv@chemicon.com www.chemicon.com
Costar	1 Alewife Center; Cambridge MA 02140, USA	1 (617) 868-6200	1 (617) 868-2076	
Cyberlab, Inc.	36 Del Mar Drive, Brookfield, CT 06804, USA	1 (203) 740-3565	1 (203) 740-3566	www.cyber-lab.com
D.R.O.P.	Z.I. la Gloriette; R.N. 92 – Chatte, F-38160 St. Marcellin, France	+33 (0)4 76 64 91 08	+33 (0)4 76 64 91 07	
Douglas Instruments	25J Thames House, 140 Battersea Park Rd., London SW11 4NB, UK	+44 (0) 171 738 8226	+44 (0) 171 738 8227	info@douglas.co.uk www.douglas.co.uk

Continued on next page

Company	Address	Telephone	Fax	E-mail, Website
EFD	977 Waterman Ave., East Providence, RI 02914, USA	1 (401) 434-1680	1 (401) 431-0237	
EFD International Inc	Unit 14, Apex Business Centre, Boscombe Rd, Dunstable, Bedfordshire LU5 4SB, UK	+44 (0)1582 666 334 Free phone: 0800 585 733 (UK) 0800 8272 9444 (Ireland)	+44 (0)1582 664 227	
Emerald BioStructures, Inc.	7869 N.E Day Rd. W., Bainbridge Island, WA 98110, USA	1 (206) 780-8535	1 (206) 780 8549	info@emeraldbiostructures.com www.emeraldbiostructures.com
Eppendorf-Netheler-Hinz GmbH	D-22331 Hamburg, Germany	+49 040 5 3801 0	+49 040 5 3801 556	eppendorf@eppendorf.com www.eppendorf.com
Falcon	Becton Dickinson Labware, 2 Oak Park, Bedford, MA 01730-9902, USA	1 (781) 275-0004	1 (781) 275-0043	mail@cbpi.com
Fluka Chemie AG	Industriestrasse 25, CH-9471 Buchs, Switzerland	+41 81 755 2511 contact your local Sigma-Aldrich supplier	+41 81 756 5449	fluka@sial.com see: www.sigma-aldrich.com

Continued on next page

Company	Address	Telephone	Fax	E-mail, Website
Gilson	72 rue Gambetta B.P. 45-95400 Villiers-le-Bel, France	+33 1 34 29 50 00	+33 1 34 29 50 80	gilson@coil.com
Glas Technik & Konstruktion	Reiherallee 12, 13503 Berlin, Germany	+49 30 4 31 61 72	+49 30 4 36 26 83	
Hamilton Bonaduz AG	P O Box 26, CH-7402, Bonaduz, Switzerland	+41 81 37 01 01 +41 81 660 6060	+41 81 37 25 63 +41 81 660 6070	hamopd@bonaduz. hamilton.ch
Hamilton Instruments (USA)	P O Box 10030, Reno, NV 89520, USA	1 (702) 858-3000 toll free US: 1-800-648-5950	1 (702) 856-7259	sales@hamiltoncomp.com www.hamiltoncomp.com
Hampton Research	27632 El Lazo Rd., Suite 100, Laguna Niguel, CA 92677-3913, USA	1 (949) 425 1321	1 (949) 425-1611	info@hrmail.com www.hamptonresearch.com
Huntsman Corp.	3040 Post Oak Blvd., Houston, TX 77056, USA	1 (713) 235-6000	1 (713) 235-6437	
ICN Biomedicals, Inc.	3300 Hyland Ave., Costa Mesa, CA 92626, USA	1 (714) 545-0100	1 (714) 557-4872	biomark@icnbiomed.com www.icnbiomed.com
Merck KGaA	D-64271 Darmstadt, Germany	+49 61 51 72 0	+49 61 5172 2000	

Continued on next page

Company	Address	Telephone	Fax	E-mail, Website
Mettler–Toledo AG	P O Box VI-400, CH-8606 Greifensee, Switzerland	+41 (81) (1) 944 22 11 +41 (1) 94445-45	+41 (81) (1) 944 31 70 +41 (1) 94445-10	order@millipore.com www.millipore.com
Millipore Intertech	P O Box 255, Bedford MA 01730, USA	1 (617) 275-9200	1 (617) 533-8630	order@millipore.com www.millipore.com
Molecular Dimensions Ltd.	61-63 Dudley Street, Luton, Beds LU2 ONP, UK	+44 1582 481884	+44 1582 481895	info@m-d-l.demon.co.uk www.m-d-l.demon.co.uk
Novus Molecular, Inc.	4660 La Jolla Village Drive, Suite 500, San Diego, CA 92122, USA	1 (619) 625-4608	1 (619) 625-4603	novus@iinet.com
Nunc AS	P O Box 280, DK4000 Roskilde, Denmark	+45 4635 9065	+45 4635 0105	
Nunc, Inc.	PO Box 20365, Rochester, NY 14602, USA	1 (716) 586-8800	1 (716) 264-3713	
Omega Technologies (European office)	P O Box 1, Broughton Astley, Leicestershire LE9 6XR, UK	+44 455 285520	+44 455 283912	
Omega Engineering, Inc. (USA)	One Omega Drive, Box 4047, Stamford, CT 06907-0047, USA	1 (203) 359-1660 toll free US: 1-800-826-6342	1 (203) 359-7700	info@omega.com intsales@omega.com www.omega.com

Continued on next page

Company	Address	Telephone	Fax	E-mail, Website
Pall Filtron Corp.	50 Bearfoot Rd., Northborough, MA 01532, USA	1 (508) 393-1800 toll free US: 1-800-FILTRON	1 (508) 393-1874	www.pall.com
Pfaltz & Bauer	172 E. Aurora Street, Waterbury, CT 06708, USA	1 (203) 574-0075	1 (203) 574-3181	www.pfaltzandbauer.com
Pharmacia Biotech	P O Box 776, 191 27 Sollentuna, Sweden	+46 8 623 8500	+46 8 623 0069	cust.servse@eu.pnu.com
Pierce Chemical Co.	P O Box 117, Rockford, IL 61105, USA	1 (815) 968-0747 toll free US: 1-800-874-3723	1 (815) 968-7316 1 (815) 968-8148 toll free US: 1-800-842-5007	cs@piercenet.com
Polycrystal	P O Box 3439; Dayton, Ohio 45401-3439, USA	1 (937) 223-9070	1 (937) 223-9070	polybook@dnaco.net
Protein Solutions Inc.	1224 West Main St., Suite 777, Charlottesville, VA 22903, USA	1 (804) 817-7177	1 (804) 817-7178	70651.3674@compuserve.com www.protein-solutions.com
Radiometer	Rue d'Alsace 72, Villeurbanne CEDEX 69627 Lyon, France	+33 47 803 3838	+33 47 868 8812	
Scientific Industries, Inc.	70 Orville Dr., Bohemia, NY 11716, USA	1 (516) 567-4700	1 (516) 567-5896	info@scind.com www.scind.com

Continued on next page

Company	Address	Telephone	Fax	E-mail, Website
Serva Feinbiochemica GmbH & Co. KG	Boehring Ingelheim Bioproducts, P O Box 105260, D-69042 Heidelberg, Germany	+49 6221 5020	+49 6221 502113	
Sialomed, Inc.	8980F Route 108, Columbia Center, Columbia, MD 21045, USA	1 (410) 997-0100	1 (410) 997-7104	
Sigma	P O Box 14508, St. Louis, MO 63178-9916, USA	1 (314) 771-5750	1 (314) 771-5757	sigma@sial.com see: www.sigma-aldrich.com
Socorex	Socorex ISBA S.A. CH-1020 Renens, Switzerland	+41 21 634 26 72	+41 21 634 27 83	socorex@socorex.ch www.socorex.ch
Whatman International Ltd.	Whatman House St. Leonard's Road, 20/20 Maidstone, Kent ME16 0LS, UK	+44 1622 676670	+44 1622 677011	information@whatman.co.uk www.whatman.com
World Precision Instruments	International Trade Center, 175 Sarasota Center Blvd., Sarasota, FL 34240-9258, USA	1 (941) 371-1003	1 (941) 377-5428	wpi@wpiinc.com www.wpiinc.com

Continued on next page

Company	Address	Telephone	Fax	E-mail, Website
World Precision Instruments (UK) Ltd	Astonbury Farm Business Centre, Aston, Stevenage, Hertfordshire SG2 7EG, UK	+44 1438 880025	+44 1438 880026	wpi@wpiuk.demon.co.uk
Wyatt Technology Corp.	30 South La Patera Lane, B-7, Santa Barbara, CA 93117, USA	1 (805) 681-9009	1 (805) 681-0123	info@wyatt.com www.wyatt.com

Useful Websites

Kerstin Fridborg

Uppsala University, Uppsala, Sweden

1. http://ibm4.carb.nist.gov:4400/

Biological Macromolecule Crystallization Database (BMCD)
An extremely useful source of statistics and information about previously crystallized biological macromolecules. Lists 2,218 crystal forms for 1,465 biological macromolecules. It can be used for comparing crystallization conditions for homologous proteins or families of proteins, designing screens, finding references, heavy atom searches, etc.

References:

1. Gilliland GL, Tung M, Blakeslee DM, Ladner J: **The biological macromolecule crystallization database, version 3.0: new features, data, and the NASA archive for protein crystal growth data.** *Acta Cryst* 1994, **D50**:408-413.

2. Gilliland G: **Biological macromolecule crystallization database.** *Meth Enzymol* 1997, **277**:546-556.

2. http://www.hamptonresearch.com/index.html

Hampton Research Company's website
Contains a wealth of practical hints and there is a catalogue of their products, useful for crystallization. The link "Info & Ideas" gives a lot of know-how. Lists all crystallization references year by year. Excellent source of information.

3. http://bmbsgi13.leeds.ac.uk/wwwprg/stura/cryst.html

Enrico Stura's website
Temperature effects, streak seeding, macroseeding, screening, additives, proteins that bind phosphate moieties, detergents for protein crystallization, protein heterogeneity and its effects on resolution, epitaxial seeding.

4. http://www.biophys.mpg.de/kuehlbrandt/zeelen/xtal.html

Johan Zeelen's website
Homogeneity of the protein sample, crystallization methods, equipment and other tools, screening for crystals, description of what you see in the drop, mounting crystals.

5. http://alpha2.bmc.uu.se/terese

Terese Bergfors' home page
Pictorial library of crystallization drop phenomena; new tips and gizmos for crystallization; yearly course in "Practical Protein Crystallization" at Uppsala University, Sweden.

6. http://wserv1.dl.ac.uk/CCP/CCP14/ccp/web-mirrors/llnlrupp/

Brent Segelke's website
CRYSTOOL program for generating your own random crystallization screen. Computing of a high efficiency screening protocol for crystals. Comparison of three screening methods for their efficiency in crystallizing five test proteins. (Random sampling was best.)

7. http://www-structure.llnl.gov/crystal_lab/cystalmake.html

A tutorial for how to grow protein crystals; useful tips and links.

8. These four all deal with heavy atom derivatives.

http://bmbsgi13.leeds.ac.uk/wwwprg/stura/heavy.html
http://www.imb-jena.de/www_sbx/thomas/manch/heavy2.html
http://bonsai.lif.icnet.uk/bmm/had/heavyatom.html
http://lsx12e.nsls.bnl.gov/x12c/mr_table.html

9. http://www.emeraldbiostructures.com

Emerald BioStructures, in collaboration with S. Safarty and W. Hol at the University of Washington, markets the following screens: Cryo I and II, Wizard I and II (U.S. Patents Pending). See appendix A5.

10. http://www.protein-solutions.com

Protein Solutions, Inc.
Lists references that mention dynamic light scattering for protein crystallization.

11. http://www.chemistry.ucsc.edu/teaching/Chem200A/ch200ref2.html

References on protein aggregation; protein solubility.

12. http://www.huntsman.com

Manufacturer and supplier of Jeffamine.

13. http://wild-turkey.mit.edu/chemicool
http://www.shef.ac.uk/~chem/web-elements

Interactive access to the periodic table with information about every element.

14. http://chemfinder.camsoft.com

Chem Finder service give access to many relevant web resources for "small molecules", e.g., physical data, hazardous materials, carcinogenicity, suppliers.

15. http://www4.ncbi.nlm.nih.gov/PubMed/

Access to the nine million citations on MedLine and more.

16. http://research.nwfsc.noaa.gov/msds.html

Material safety data searches.

17. http://www.hhmi.org/science/labsafe/lcss/start.htm

Laboratory chemical safety summaries.

18. http://www.geocities.com/Athens/Forum/7504/xtaljokes.html

The Crystallography Jokes Home Page (if you are bored while waiting for your crystals to grow).

Appendix A5

Commercially Available Screens

A5.1.

Screen name: Crystal Screen I
Supplier: Hampton Research (reprinted with permission)

1	30% MPD	0.1 M Na acetate pH 4.6	0.02 M Ca chloride
2	0.4 M K Na tartrate	none	none
3	0.4 M ammonium phosphate	none	none
4	2.0 M ammonium sulfate	0.1 M Tris HCl pH 8.5	none
5	30% MPD	0.1 M Na HEPES pH 7.5	0.2 M Na citrate
6	30% PEG 4000	0.1 M Tris HCl pH 8.5	0.2 M Mg chloride
7	1.4 M Na acetate	0.1 M Na cacodylate pH 6.5	none
8	30% 2-propanol	0.1 M Na cacodylate pH 6.5	0.2 M Na citrate
9	30% PEG 4000	0.1 M Na citrate pH 5.6	0.2 M ammonium acetate
10	30% PEG 4000	0.1 M Na acetate pH 4.6	0.2 M ammonium acetate
11	1.0 M ammonium phosphate	0.1 M Na citrate pH 5.6	none

Continued on next page

12	30% 2-propanol	0.1 M Na HEPES pH 7.5	0.2 M Mg chloride
13	30% PEG 400	0.1 M Tris HCl pH 8.5	0.2 M Na citrate
14	28% PEG 400	0.1 M Na HEPES pH 7.5	0.2 M Ca chloride
15	30% PEG 8000	0.1 M Na cacodylate pH 6.5	0.2M ammonium sulfate
16	1.5 M Li sulfate	0.1 M Na HEPES pH 7.5	none
17	30% PEG 4000	0.1 M Tris HCl pH 8.5	0.2 M Li sulfate
18	20% PEG 8000	0.1 M Na cacodylate pH 6.5	0.2 M Mg acetate
19	30% 2-propanol	0.1 M Tris HCl pH 8.5	0.2 M ammonium acetate
20	25% PEG 4000	0.1 M Na acetate pH 4.6	0.2 M ammonium sulfate
21	30% MPD	0.1 M Na cacodylate pH 6.5	0.2 M Mg acetate
22	30% PEG 4000	0.1 M Tris HCl pH 8.5	0.2 M Na acetate
23	30% PEG 400	0.1 M Na HEPES pH 7.5	0.2 M Mg chloride
24	20% 2-propanol	0.1 M Na acetate pH 4.6	0.2 M Ca chloride
25	1.0 M Na acetate	0.1 M imidazole pH 6.5	none
26	30% MPD	0.1 M Na citrate pH 5.6	0.2 M ammonium acetate
27	20% 2-propanol	0.1 M Na HEPES pH 7.5	0.2 M Na citrate
28	30% PEG 8000	0.1 M Na cacodylate pH 6.5	0.2 M Na acetate
29	0.8 M K Na tartrate	0.1 M Na HEPES pH 7.5	none
30	30% PEG 8000	none	0.2 M ammonium sulfate
31	30% PEG 4000	none	0.2 M ammonium sulfate
32	2.0 M ammonium sulfate	none	none
33	4.0 M Na formate	none	none

Continued on next page

34	2.0 M Na formate	0.1 M Na acetate pH 4.6	none
35	1.6 M Na/K phosphate	0.1 M Na HEPES pH 7.5	none
36	8% PEG 8000	0.1 M Tris HCl pH 8.5	none
37	8% PEG 4000	0.1 M Na acetate pH 4.6	none
38	1.4 M Na citrate	0.1 M Na hepes pH 7.5	none
39	2% PEG 400	0.1 M Na HEPES pH 7.5	2.0 M ammonium sulfate
40	20% 2-propanol	0.1 M Na citrate pH 5.6	20% PEG 4000
41	10% 2-propanol	0.1 M Na HEPES pH 7.5	20% PEG 4000
42	20% PEG 8000	none	0.05 M K phosphate
43	30% PEG 1500	none	none
44	0.2 M Mg formate	none	none
45	18% PEG 8000	0.1 M Na cacodylate pH 6.5	0.2 M Zn acetate
46	18% PEG 8000	0.1 M Na cacodylate pH 6.5	0.2 M Ca acetate
47	2.0 M ammonium sulfate	0.1 M Na acetate pH 4.6	none
48	2.0 M ammonium phosphate	0.1 M Tris HCl pH 8.5	none
49	2% PEG 8000	none	1.0 M Li sulfate
50	15% PEG 8000	none	0.5 M Li sulfate

A5.2.

Screen name: Crystal Screen II
Supplier: Hampton Research (reprinted with permission)

#			
1	10% PEG 6000	none	2.0 M Na chloride
2	0.5 M NaCl	0.01 M CTAB	0.01 M Mg chloride
3	25% ethylene glycol	none	none
4	35% dioxane	none	none
5	5% isopropanol	none	2.0 M ammonium sulfate
6	1.0 M imidazole pH 7.0	none	none
7	10% PEG 1000	none	10% PEG 8000
8	10% ethanol	none	1.5 M Na chloride
9	2.0 M Na chloride	0.1 M Na acetate pH 4.6	none
10	30% MPD	0.1 M Na acetate pH 4.6	0.2 M NaCl
11	1.0 M 1,6-hexanediol	0.1 M Na acetate pH 4.6	0.01 M Co chloride
12	30% PEG 400	0.1 M Na acetate pH 4.6	0.1 M Cd chloride
13	30% PEG MME 2000	0.1 M Na acetate pH 4.6	0.2 M ammonium sulfate
14	2.0 M ammonium sulfate	0.1M Na citrate pH 5.6	0.2 M K/Na tartrate
15	1.0 M Li sulfate	0.1M Na citrate pH 5.6	0.5 M ammonium sulfate

Continued on next page

16	2% polyethyleneimine	0.1 M Na citrate pH 5.6	0.5 M Na chloride
17	35% *tert*-butanol	0.1 M Na citrate pH 5.6	none
18	10% Jeffamine M-600	0.1 M Na citrate pH 5.6	0.01 M ferric chloride
19	2.5 M 1,6-hexanediol	0.1 M Na citrate pH 5.6	none
20	1.6 M Mg sulfate	0.1 M MES pH 6.5	none
21	2.0 M Na chloride	0.1 M MES pH 6.5	0.2 M Na/K phosphate
22	12% PEG 2000	0.1 M MES pH 6.5	none
23	10% dioxane	0.1 M MES pH 6.5	1.6 M ammonium sulfate
24	30% Jeffamine M-600	0.1 M MES pH 6.5	0.05 M Cs chloride
25	1.8 M ammonium sulfate	0.1 M MES pH 6.5	0.01 M Co chloride
26	30% PEG MME 5000	0.1 M MES pH 6.5	0.2 M ammonium sulfate
27	25% PEG MME 550	0.1 M MES pH 6.5	0.01 M Zn sulfate
28	1.6 M Na citrate pH 6.5	none	none
29	30% MPD	0.1 M HEPES pH 7.5	0.5 M ammonium sulfate
30	10% PEG 6000	0.1 M HEPES pH 7.5	5% MPD
31	20% Jeffamine M-600	0.1 M HEPES pH 7.5	none
32	1.6 M ammonium sulfate	0.1 M HEPES pH 7.5	0.1 M Na chloride
33	2.0 M ammonium formate	0.1 M HEPES pH 7.5	none
34	1.0 M Na acetate	0.1 M HEPES pH 7.5	0.05 M Cd sulfate
35	70% MPD	0.1 M HEPES pH 7.5	none
36	4.3 M Na chloride	0.1 M HEPES pH 7.5	none

Continued on next page

37	10% PEG 8000	0.1 M HEPES pH 7.5	8% ethylene glycol
38	20% PEG 1000	0.1 M HEPES pH 7.5	none
39	3.4 M 1,6-hexanediol	0.1 M Tris pH 8.5	0.2 M Mg chloride
40	25% *tert*-butanol	0.1 M Tris pH 8.5	0.1 M Ca chloride
41	1.0 M Li sulfate	0.1 M Tris pH 8.5	0.01 M Ni chloride
42	12% glycerol	0.1 M Tris pH 8.5	1.5 M ammonium sulfate
43	50% MPD	0.1 M Tris pH 8.5	0.2 M ammonium phosphate
44	20% ethanol	0.1 M Tris pH 8.5	none
45	20% PEG MME 2000	0.1 M Tris pH 8.5	0.01 M Ni chloride
46	30% PEG MME 550	0.1 M bicine pH 9.0	0.1 M Na chloride
47	2.0 M Mg chloride	0.1 M bicine pH 9.0	none
48	10% PEG 2000	0.1 M bicine pH 9.0	2% dioxane

A5.3.

Screen name: **Additive Screens 1, 2, and 3**
Supplier: Hampton Research (reprinted with permission)

In addition, Hampton Research has many other screens not listed here; see their website or catalogue.

Additive Screen 1 Formulation:

1.	0.1 M barium chloride	Cation
2.	0.1 M cadmium chloride	Cation
3.	0.1 M calcium chloride	Cation
4.	0.1 M cobalt chloride	Cation
5.	0.1 M copper chloride	Cation
6.	0.1 M magnesium chloride	Cation
7.	0.1 M manganese chloride	Cation
8.	0.1 M strontium chloride	Cation
9.	0.1 M ytrium chloride	Cation
10.	0.1 M zinc chloride	Cation
11.	30% dioxane	Organic
12.	30% ethanol	Organic
13.	30% ethylene glycol	Organic
14.	30% glycerol	Organic
15.	30% 1,6-hexanediol	Organic
16.	30% isopropanol	Organic
17.	30% methanol	Organic
18.	30% 2-methyl-2,4-pentanediol	Organic
19.	50% PEG 400	Organic
20.	0.1 M trimethylamine HCl	Chaotrope
21.	1.0 M guanidine HCl	Chaotrope
22.	0.1 M urea	Chaotrope
23.	15% 1,2,3-heptanetriol	Amphiphile
24.	20% benzamidine HCl	Amphiphile

Additive Screen 2 Formulation:

1.	1.0 M sodium iodide	Ion
2.	0.1 M L-cysteine	Reducing agent
3.	0.1 M EDTA, sodium salt	Chelator

Continued on next page

4.	0.1 M b-nicotinamide adenine dinucleotide	Co-factor
5.	0.1 M adenosine-51-triphosphate disodium salt	Co-factor
6.	30% w/v D(+)-glucose	Carbohydrate
7.	30% w/v D(+)-sucrose	Carbohydrate
8.	30% w/v xylitol	Carbohydrate
9.	0.1 M spermidine	Polyamine
10.	0.1 M spermine-tetrahydrochloride	Polyamine
11.	30% w/v 6-aminocaproic acid	Linker
12.	30% w/v 1,5-diaminopentane dihydrochloride	Linker
13.	30% v/v 1,6-diaminohexane	Linker
14.	30% v/v 1,8-diaminooctane	Linker
15.	1.0 M glycine	Linker
16.	0.1 M glycyl-glycyl-glycine	Linker
17.	0.1 M hexaminecobalt trichloride	Polyamine
18.	0.1 M taurine	Linker
19.	0.1 M betaine monohydrate	Linker
20.	5% w/v polyvinylpyrrolidone K15	Polymer
21.	3.0 M non-detergent sulfo-betaine 195	Solubilizing agent
22.	2.0 M non-detergent sulfo-betaine 201	Solubilizing agent
23.	30% v/v dimethyl sulfoxide	Dissociating agent
24.	0.1 M phenol	Chaotrope

Additive Screen 3 Formulation:

1.	1.0 M ammonium sulfate	Salt
2.	1.0 M cesium chloride	Salt
3.	1.0 M potassium chloride	Salt
4.	1.0 M lithium chloride	Salt
5.	2.0 M sodium chloride	Salt
6.	0.5 M sodium fluoride	Salt
7.	2.0 M sodium thiocyanate	Salt
8.	30% w/v dextran sulfate, sodium salt	Polymer
9.	50% v/v Jeffamine M-600	Organic, non-volatile
10.	40% v/v 2,5-hexanediol	Organic, non-volatile
11.	40% v/v 1,3-butanediol	Organic, non-volatile
12.	40% v/v polypropylene glycol P400	Organic, non-volatile
13.	40% v/v 1,4-butanediol	Organic, non-volatile
14.	40% v/v *tert*-butanol	Organic, volatile
15.	40% v/v 1,3-propanediol	Organic, volatile

Continued on next page

16.	40% v/v acetonitrile	Organic, volatile
17.	40% v/v γ-butyrolactone	Organic, volatile
18.	40% v/v n-propanol	Organic, volatile
19.	5% v/v ethyl acetate	Organic, volatile
20.	40% v/v acetone	Organic, volatile
21.	2.5% v/v dichloromethane	Organic, volatile
22.	7% v/v n-butanol	Organic, volatile
23.	40% v/v 2,2,2-trifluoroethanol	Organic, volatile
24.	0.1 M 1,4-dithio-DL-threitol	Reducing agent

A5.4.

Screen name: Wizard I
Supplier: Emerald BioStructures, Inc. (reprinted with permission)*

	precipitant	buffer	salt
1	20% (w/v) PEG-8000	0.1 M CHES pH 9.5	none
2	10% (v/v) 2-propanol	0.1 M HEPES pH 7.5	0.2 M NaCl
3	15% (v/v) ethanol	0.1 M CHES pH 9.5	none
4	35% (v/v) 2-methyl-2,4-pentanediol	0.1 M imidazole pH 8.0	0.2 M MgCl$_2$
5	30% (v/v) PEG-400	0.1 M CAPS pH 10.5	none
6	20% (w/v) PEG-3000	0.1 M citrate pH 5.5	none
7	10% (w/v) PEG-8000	0.1 M MES pH 6.0	0.2 M Zn(OAc)$_2$
8	2.0 M (NH$_4$)$_2$SO$_4$	0.1 M citrate pH 5.5	none
9	1.0 M (NH$_4$)$_2$HPO$_4$	0.1 M acetate pH 4.5	none
10	20% (w/v) PEG-2000 MME	0.1 M Tris pH 7.0	none
11	20% (v/v) 1,4-butanediol	0.1 M MES pH 6.0	0.2 M Li$_2$SO$_4$
12	20% (w/v) PEG-1000	0.1 M imidazole pH 8.0	0.2 M Ca(OAc)$_2$
13	1.26 M (NH$_4$)$_2$SO$_4$	0.1 M cacodylate pH 6.5	none
14	1.0 M sodium citrate	0.1 M cacodylate pH 6.5	none
15	10% (w/v) PEG-3000	0.1 M imidazole pH 8.0	0.2 M Li$_2$SO$_4$

Continued on next page

	precipitant	buffer	salt
16	2.5 M NaCl	0.1 M Na/K phosphate pH 6.2	none
17	30% (w/v) PEG-8000	0.1 M acetate pH 4.5	0.2 M Li$_2$SO$_4$
18	1.0 M K/Na tartrate	0.1 M imidazole pH 8.0	0.2 M NaCl
19	20% (w/v) PEG-1000	0.1 M Tris pH 7.0	none
20	0.4 M NaH$_2$PO$_4$/1.6 M K$_2$HPO$_4$	0.1 M imidazole pH 8.0	0.2 M NaCl
21	20% (w/v) PEG-8000	0.1 M HEPES pH 7.5	none
22	10% (v/v) 2-propanol	0.1 M Tris pH 8.5	none
23	15% (v/v) ethanol	0.1 M imidazole pH 8.0	0.2 M MgCl$_2$
24	35% (v/v) 2-methyl-2,4-pentanediol	0.1 M Tris pH 7.0	0.2 M NaCl
25	30% (v/v) PEG-400	0.1 M Tris pH 8.5	0.2 M MgCl$_2$
26	10% (w/v) PEG-3000	0.1 M CHES pH 9.5	none
27	1.2 M NaH$_2$PO$_4$/0.8 M K$_2$HPO$_4$	0.1 M CAPS pH 10.5	0.2 M Li$_2$SO$_4$
28	20% (w/v) PEG-3000	0.1 M HEPES pH 7.5	0.2 M NaCl
29	10% (w/v) PEG-8000	0.1 M CHES pH 9.5	0.2 M NaCl
30	1.26 M (NH$_4$)$_2$SO$_4$	0.1 M acetate pH 4.5	0.2 M NaCl
31	20% (w/v) PEG-8000	0.1 M phosphate-citrate pH 4.2	0.2 M NaCl
32	10% (w/v) PEG-3000	0.1 M Na/K phosphate pH 6.2	none
33	2.0 M (NH$_4$)$_2$SO$_4$	0.1 M CAPS pH 10.5	0.2 M Li$_2$SO$_4$
34	1.0 M (NH$_4$)$_2$HPO$_4$	0.1 M imidazole pH 8.0	none
35	20% (v/v) 1,4-butanediol	0.1 M acetate pH 4.5	none

Continued on next page

	precipitant	buffer	salt
36	1.0 M sodium citrate	0.1 M imidazole pH 8.0	none
37	2.5 M NaCl	0.1 M imidazole pH 8.0	none
38	1.0 M K/Na tartrate	0.1 M CHES pH 9.5	0.2 M Li_2SO_4
39	20% (w/v) PEG-1000	0.1 M phosphate-citrate pH 4.2	0.2 M Li_2SO_4
40	10% (v/v) 2-propanol	0.1 M MES pH 6.0	0.2 M $Ca(OAc)_2$
41	30% (w/v) PEG-3000	0.1 M CHES pH 9.5	none
42	15% (v/v) ethanol	0.1 M Tris pH 7.0	none
43	35% (v/v) 2-methyl-2,4-pentanediol	0.1 M Na/K phosphate pH 6.2	none
44	30% (v/v) PEG-400	0.1 M acetate pH 4.5	0.2 M $Ca(OAc)_2$
45	20% (w/v) PEG-3000	0.1 M acetate pH 4.5	none
46	10% (w/v) PEG-8000	0.1 M imidazole pH 8.0	0.2 M $Ca(OAc)_2$
47	1.26 M $(NH_4)_2SO_4$	0.1 M Tris pH 8.5	0.2 M Li_2SO_4
48	20% (w/v) PEG-1000	0.1 M acetate pH 4.5	0.2 M $Zn(OAc)_2$

* The Crystal Growth Matrices were developed by Steve L. Sarfaty and Wim G. J. Hol at the University of Washington (Seattle, WA, USA, 1998). Emerald BioStructures, Inc., has obtained an exclusive license from the University of Washington to market the Crystal Growth Matrices.

A5.5.

Screen name: Wizard II

Supplier: Emerald BioStructures, Inc. (reprinted with permission)*

	precipitant	buffer	salt
1	10% (w/v) PEG-3000	0.1 M acetate pH 4.5	0.2 M Zn(OAc)$_2$
2	35% (v/v) 2-methyl-2.4-pentadiol	0.1 M MES pH 6.0	0.2 M Li$_2$SO$_4$
3	20% (w/v) PEG-8000	0.1 M Tris pH 8.5	0.2 M MgCl$_2$
4	2.0 M (NH$_4$)$_2$SO$_4$	0.1 M cacodylate pH 6.5	0.2 M NaCl
5	20% (v/v) 1,4-butanediol	0.1 M HEPES pH 7.5	0.2 M NaCl
6	10% (v/v) 2-propanol	0.1 M phosphate-citrate pH 4.2	0.2 M Li$_2$SO$_4$
7	30% (w/v) PEG-3000	0.1 M Tris pH 7.0	0.2 M NaCl
8	10% (w/v) PEG-8000	0.1 M Na/K phosphate pH 6.2	0.2 M NaCl
9	2.0 M (NH$_4$)$_2$SO$_4$	0.1 M phosphate-citrate pH 4.2	none
10	1.0 M (NH$_4$)$_2$HPO$_4$	0.1 M Tris pH 8.5	none
11	10% (v/v) 2-propanol	0.1 M cacodylate pH 6.5	0.2 M Zn(OAc)$_2$
12	30% (v/v) PEG-400	0.1 M cacodylate pH 6.5	0.2 M Li$_2$SO$_4$
13	15% (v/v) ethanol	0.1 M citrate pH 5.5	0.2 M Li$_2$SO$_4$
14	20% (w/v) PEG-1000	0.1 M Na/K phosphate pH 6.2	0.2 M NaCl
15	1.26 M (NH$_4$)$_2$SO$_4$	0.1 M HEPES pH 7.5	none

Continued on next page

	precipitant	buffer	salt
16	1.0 M sodium citrate	0.1 M CHES pH 9.5	none
17	2.5 M NaCl	0.1 M Tris pH 7.0	0.2 M MgCl$_2$
18	20% (w/v) PEG-3000	0.1 M Tris pH 7.0	0.2 M Ca(OAc)$_2$
19	1.6 M NaH$_2$PO$_4$/0.4 M K$_2$HPO$_4$	0.1 M phosphate-citrate pH 4.2	none
20	15% (v/v) ethanol	0.1 M MES pH 6.0	0.2 M Zn(OAc)$_2$
21	35% (v/v) 2-methyl-2,4-pentadiol	0.1 M acetate pH 4.5	none
22	10% (v/v) 2-propanol	0.1 M imidazole pH 8.0	none
23	15% (v/v) ethanol	0.1 M HEPES pH 7.5	0.2 M MgCl$_2$
24	30% (w/v) PEG-8000	0.1 M imidazole pH 8.0	0.2 M NaCl
25	35% (v/v) 2-methyl-2,4-pentadiol	0.1 M HEPES pH 7.5	0.2 M NaCl
26	30% (w/v) PEG-400	0.1 M CHES pH 9.5	none
27	10% (w/v) PEG-3000	0.1 M cacodylate pH 6.5	0.2 M MgCl$_2$
28	20% (w/v) PEG-8000	0.1 M MES pH 6.0	0.2 M Ca(OAc)$_2$
29	1.26 M (NH$_4$)$_2$SO$_4$	0.1 M CHES pH 9.5	0.2 M NaCl
30	20% (v/v) 1,4-butanediol	0.1 M imidazole pH 8.0	0.2 M Zn(OAc)$_2$
31	1.0 M sodium citrate	0.1 M Tris pH 7.0	0.2 M NaCl
32	20% (w/v) PEG-1000	0.1 M Tris pH 8.5	none
33	1.0 M (NH$_4$)$_2$HPO$_4$	0.1 M citrate pH 5.5	0.2 M NaCl
34	10% (w/v) PEG-8000	0.1 M imidazole pH 8.0	none
35	0.8 M NaH$_2$PO$_4$/1.2 M K$_2$HPO$_4$	0.1 M acetate pH 4.5	none

Continued on next page

	precipitant	buffer	salt
36	10% (w/v) PEG-3000	0.1 M phosphate-citrate pH 4.2	0.2 M NaCl
37	1.0 M K/Na tartrate	0.1 M Tris pH 7.0	0.2 M Li_2SO_4
38	2.5 M NaCl	0.1 M acetate pH 4.5	0.2 M Li_2SO_4
39	20% (w/v) PEG-8000	0.1 M CAPS pH 10.5	0.2 M NaCl
40	20% (w/v) PEG-3000	0.1 M imidazole pH 8.0	0.2 M $Zn(OAc)_2$
41	2.0 M $(NH_4)_2SO_4$	0.1 M Tris pH 7.0	0.2 M Li_2SO_4
42	30% (v/v) PEG-400	0.1 M HEPES pH 7.5	0.2 M NaCl
43	10% (w/v) PEG-8000	0.1 M Tris pH 7.0	0.2 M $MgCl_2$
44	20% (w/v) PEG-1000	0.1 M cacodylate pH 6.5	0.2 M $MgCl_2$
45	1.26 M $(NH_4)_2SO_4$	0.1 M MES pH 6.0	none
46	1.0 M $(NH_4)_2HPO_4$	0.1 M imidazole pH 8.0	0.2 M NaCl
47	2.5 M NaCl	0.1 M imidazole pH 8.0	0.2 M $Zn(OAc)_2$
48	1.0 M K/Na tartrate	0.1 M MES pH 6.0	none

* The Crystal Growth Matrices were developed by Steve L. Sarfaty and Wim G. J. Hol at the University of Washington (Seattle, WA, USA, 1998). Emerald BioStructures, Inc., has obtained an exclusive license from the University of Washington to market the Crystal Growth Matrices.

A5.6.

Screen name: Cryo I
Supplier: Emerald BioStructures, Inc. (reprinted with permission)*

	precipitant	buffer	additive(s)
1	40% (v/v) 2-methyl-2,4-pentanediol	0.1 M phosphate-citrate pH 4.2	none
2	40% (v/v) ethylene glycol	0.1 M acetate pH 4.5	none
3	50% (v/v) PEG-200	0.1 M citrate pH 5.5	none
4	40% (v/v) PEG-300	0.1 M HEPES pH 7.5	0.2 M NaCl
5	40% (v/v) PEG-400	0.1 M citrate pH 5.5	0.2 M MgCl$_2$
6	40% (v/v) PEG-600	0.1 M cacodylate pH 6.5	0.2 M Ca(OAc)$_2$
7	40% (v/v) ethanol	0.1 M Tris pH 8.5	0.05 M MgCl$_2$
8	35% (v/v) 2-ethoxyethanol	0.1 M cacodylate pH 6.5	none
9	35% (v/v) 2-propanol	0.1 M phosphate-citrate pH 4.2	none
10	45% (v/v) glycerol	0.1 M imidazole pH 8.0	none
11	35% (v/v) 2-methyl-2,4-pentanediol	0.1 M Tris pH 8.5	0.2 M (NH$_4$)$_2$SO$_4$
12	50% (v/v) ethylene glycol	0.1 M acetate pH 4.5	5% (w/v) PEG-1000
13	30% (v/v) PEG-200	0.1 M MES pH 6.0	5% (w/v) PEG-3000
14	20% (v/v) PEG-300	0.1 M phosphate-citrate pH 4.2	0.2 M (NH$_4$)$_2$SO$_4$, 10% (v/v) glycerol
15	50% (v/v) PEG-400	0.1 M CHES pH 9.5	0.2 M NaCl

Continued on next page

	precipitant	buffer	additive(s)
16	30% (v/v) PEG-600	0.1 M MES pH 6.0	5% (w/v) PEG-1000, 10% (v/v) glycerol
17	40% (v/v) 1,2-propanediol	0.1 M HEPES pH 7.5	none
18	35% (v/v) 2-ethoxyethanol	0.1 M imidazole pH 8.0	0.05 M Ca(OAc)$_2$
19	35% (v/v) 2-propanol	0.1 M Tris pH 8.5	none
20	30% (v/v) 1,2-propanediol	0.1 M citrate pH 5.5	20% (v/v) 2-methyl-2,4-pentanediol
21	40% (v/v) 1,2-propanediol	0.1 M acetate pH 4.5	0.05 M Ca(OAc)$_2$
22	40% (v/v) ethylene glycol	0.1 M Na/K phosphate pH 6.2	none
23	40% (v/v) 2-methyl-2,4-pentanediol	0.1 M Tris pH 7.0	0.2 M (NH$_4$)$_2$SO$_4$
24	40% (v/v) PEG-400	0.1 M Na/K phosphate pH 6.2	0.2 M NaCl
25	30% (v/v) PEG-200	0.1 M Tris pH 8.5	0.2 M (NH$_4$)$_2$HPO$_4$
26	40% (v/v) PEG-300	0.1 M CHES pH 9.5	0.2 M NaCl
27	30% (v/v) PEG-400	0.1 M CAPS pH 10.5	0.5 M (NH$_4$)$_2$SO$_4$, 10% (v/v) glycerol
28	30% (v/v) PEG-600	0.1 M HEPES pH 7.5	0.05 M Li$_2$SO$_4$, 10% (v/v) glycerol
29	40% (v/v) PEG-300	0.1 M CHES pH 9.5	0.2 M sodium citrate
30	35% (v/v) 2-ethoxyethanol	0.1 M citrate pH 5.5	none
31	35% (v/v) 2-propanol	0.1 M citrate pH 5.5	5% (w/v) PEG-1000
32	40% (v/v) 1,2-propanediol	0.1 M CHES pH 9.5	0.2 M sodium citrate
33	25% (v/v) 1,2-propanediol	0.1 M imidazole pH 8.0	0.2 M Zn(OAc)$_2$, 10% (v/v) glycerol
34	40% (v/v) 2-methyl-2,4-pentanediol	0.1 M imidazole pH 8.0	0.2 M MgCl$_2$
35	40% (v/v) ethylene glycol	0.1 M HEPES pH 7.5	5% (w/v) PEG-3000

Continued on next page

	precipitant	buffer	additive(s)
36	50% (v/v) PEG-200	0.1 M Tris pH 7.0	0.05 M Li$_2$SO$_4$
37	40% (v/v) PEG-300	0.1 M cacodylate pH 6.5	0.2 M Ca(OAc)$_2$
38	40% (v/v) PEG-400	0.1 M Tris pH 8.5	0.2 M Li$_2$SO$_4$
39	40% (v/v) PEG-600	0.1 M phosphate-citrate pH 4.2	none
40	40% (v/v) ethanol	0.1 M phosphate-citrate pH 4.2	5% (w/v) PEG-1000
41	25% (v/v) 1,2-propanediol	0.1 M phosphate-citrate pH 4.2	5% (w/v) PEG-3000, 10% (v/v) glycerol
42	40% (v/v) ethylene glycol	0.1 M Tris pH 7.0	none
43	50% (v/v) ethylene glycol	0.1 M Tris pH 8.5	0.2 M MgCl$_2$
44	50% (v/v) PEG-200	0.1 M cacodylate pH 6.5	0.2 M Zn(OAc)$_2$
45	20% (v/v) PEG-300	0.1 M Tris pH 8.5	5% (w/v) PEG-8000, 10% (v/v) glycerol
46	40% (v/v) PEG-400	0.1 M MES pH 6.0	5% (w/v) PEG-3000
47	50% (v/v) PEG-400	0.1 M acetate pH 4.5	0.2 M Li$_2$SO$_4$
48	40% (v/v) PEG-600	0.1 M imidazole pH 8.0	0.2 M Zn(OAc)$_2$

* The Cryo Crystal Growth Matrices were developed by Steve L. Sarfaty and Wim G. J. Hol at the University of Washington (Seattle, WA, USA, 1998), with subsequent modification by Hidong Kim and Lance J. Stewart at Emerald BioStructures, Inc. Emerald BioStructures, Inc., has obtained an exclusive license from the University of Washington to market the Cryo Crystal Growth Matrices.

A5.7.

Screen name: Cryo II
Supplier: Emerald BioStructures, Inc. (reprinted with permission)*

	precipitant	buffer	additive(s)
1	40% (v/v) 2-methyl-2,4-pentanediol	0.1 M cacodylate pH 6.5	5% (w/v) PEG-8000
2	50% (v/v) PEG-200	0.1 M CHES pH 9.5	none
3	40% (v/v) ethylene glycol	0.1 M phosphate-citrate pH 4.2	0.2 M $(NH_4)_2SO_4$
4	40% (v/v) PEG-400	0.1 M HEPES pH 7.5	0.2 M $Ca(OAc)_2$
5	40% (v/v) PEG-300	0.1 M Tris pH 7.0	5% (w/v) PEG-1000
6	30% (v/v) PEG-600	0.1 M cacodylate pH 6.5	1 M NaCl, 10% (v/v) glycerol
7	40% (v/v) ethanol	0.1 M Tris pH 7.0	none
8	35% (v/v) 2-ethoxyethanol	0.1 M Na/K phosphate pH 6.2	0.2 M NaCl
9	35% (v/v) 2-propanol	0.1 M imidazole pH 8.0	0.05 M $Zn(OAc)_2$
10	40% (v/v) 1,2-propanediol	0.1 M acetate pH 4.5	none
11	25% (v/v) 1,2-propanediol	0.1 M Na/K phosphate pH 6.2	10% (v/v) glycerol
12	40% (v/v) 1,2-propanediol	0.1 M citrate pH 5.5	0.2 M NaCl
13	35% (v/v) 2-methyl-2,4-pentanediol	0.1 M cacodylate pH 6.5	0.05 M $Zn(OAc)_2$
14	40% (v/v) ethylene glycol	0.1 M imidazole pH 8.0	0.2 M $Ca(OAc)_2$
15	50% (v/v) PEG-200	0.1 M Na/K phosphate pH 6.2	0.2 M NaCl

Continued on next page

	precipitant	buffer	additive(s)
16	20% (v/v) PEG-300	0.1 M imidazole pH 8.0	1 M $(NH_4)_2SO_4$, 10% (v/v) glycerol
17	50% (v/v) PEG-400	0.1 M MES pH 6.0	none
18	40% (v/v) PEG-300	0.1 M phosphate-citrate pH 4.2	none
19	40% (v/v) PEG-600	0.1 M acetate pH 4.5	0.2 M $MgCl_2$
20	50% (v/v) ethylene glycol	0.1 M CHES pH 9.5	0.5 M K/Na tartrate
21	35% (v/v) 2-ethoxyethanol	0.1 M Tris pH 8.5	0.2 M Li_2SO_4
22	35% (v/v) 2-propanol	0.1 M cacodylate pH 6.5	0.2 M $MgCl_2$
23	30% (v/v) 1,2-propanediol	0.1 M HEPES pH 7.5	20% (v/v) PEG-400
24	25% (v/v) 1,2-propanediol	0.1 M Tris pH 8.5	0.2 M $MgCl_2$, 10% (v/v) glycerol
25	40% (v/v) 2-methyl-2,4-pentanediol	0.1 M CAPS pH 10.5	none
26	40% (v/v) ethylene glycol	0.1 M MES pH 6.0	0.2 M $Zn(OAc)_2$
27	50% (v/v) PEG-200	0.1 M Tris pH 7.0	none
28	40% (v/v) PEG-300	0.1 M imidazole pH 8.0	0.2 M $Zn(OAc)_2$
29	30% (v/v) PEG-400	0.1 M HEPES pH 7.5	5% (w/v) PEG-3000, 10% (v/v) glycerol
30	40% (v/v) PEG-600	0.1 M citrate pH 5.5	none
31	40% (v/v) PEG-600	0.1 M CHES pH 9.5	none
32	35% (v/v) 2-propanol	0.1 M acetate pH 4.5	none
33	45% (v/v) glycerol	0.1 M cacodylate pH 6.5	0.2 M $Ca(OAc)_2$
34	25% (v/v) 1,2-propanediol	0.1 M Tris pH 7.0	0.2 M $(NH_4)_2SO_4$, 10% (v/v) glycerol
35	40% (v/v) 2-methyl-2,4-pentanediol	0.1 M citrate pH 5.5	none

Continued on next page

	precipitant	buffer	additive(s)
36	50% (v/v) PEG-200	0.1 M cacodylate pH 6.5	0.2 M MgCl$_2$
37	50% (v/v) ethylene glycol	0.1 M imidazole pH 8.0	none
38	40% (v/v) PEG-400	0.1 M acetate pH 4.5	none
39	30% (v/v) PEG-600	0.1 M Tris pH 7.0	0.5 M (NH$_4$)$_2$SO$_4$, 10% (v/v) glycerol
40	40% (v/v) 2-methyl-2,4-pentanediol	0.1 M CHES pH 9.5	none
41	50% (v/v) ethylene glycol	0.1 M HEPES pH 7.5	0.2 M Li$_2$SO$_4$
42	30% (v/v) PEG-200	0.1 M acetate pH 4.5	0.1 M NaCl
43	40% (v/v) PEG-400	0.1 M imidazole pH 8.0	none
44	35% (v/v) 2-methyl-2,4-pentanediol	0.1 M acetate pH 4.5	10% (v/v) glycerol
45	40% (v/v) PEG-300	0.1 M acetate pH 4.5	0.2 M NaCl
46	30% (v/v) PEG-200	0.1 M CAPS pH 10.5	0.2 M (NH$_4$)$_2$SO$_4$
47	50% (v/v) PEG-200	0.1 M HEPES pH 7.5	none
48	50% (v/v) PEG-200	0.1 M phosphate-citrate pH 4.2	0.2 M NaCl

* The Cryo Crystal Growth Matrices were developed by Steve L. Sarfaty and Wim G J. Hol at the University of Washington (Seattle, WA, USA, 1998), with subsequent modification by Hidong Kim and Lance J. Stewart at Emerald BioStructures, Inc. Emerald BioStructures, Inc., has obtained an exclusive license from the University of Washington to market the Cryo Crystal Growth Matrices.

A5.8.

Screen name: Structure Screen 1
Supplier: Molecular Dimension, Ltd. (reprinted with permission)

	precipitant	buffer	salt
1	0.02M calcium chloride dihydrate	0.1M Na acetate trihydrate pH 4.6	30% v/v 2-methyl-2,4 pentanediol
2	0.2M ammonium acetate	0.1M Na acetate trihydrate pH 4.6	30% w/v PEG 4000
3	0.2M ammonium sulfate	0.1M Na acetate trihydrate pH 4.6	25% w/v PEG 4000
4	none	0.1M Na acetate trihydrate pH 4.6	2.0M sodium formate
5	none	0.1M Na acetate trihydrate pH 4.6	2.0M ammonium sulfate
6	none	0.1M Na acetate trihydrate pH 4.6	8% w/v PEG 4000
7	0.2M ammonium acetate	0.1M *tri*-sodium citrate dihydrate pH 5.6	30% w/v PEG 4000
8	0.2M ammonium acetate	0.1M *tri*-sodium citrate dihydrate pH 5.6	30% v/v 2-methyl-2,4-pentanediol
9	none	0.1M *tri*-sodium citrate dihydrate pH 5.6	20% w/v 2-propanol, 20% w/v PEG 4000
10	none	0.1M Na citrate pH 5.6	1.0M ammonium dihydrogen phosphate
11	0.2M calcium chloride dihydrate	0.1M Na acetate trihydrate pH 4.6	20% v/v 2-propanol
12	none	0.1M Na cacodylate pH 6.5	1.4M Na acetate trihydrate
13	0.2M *tri*-sodium citrate dihydrate	0.1M Na cacodylate pH 6.5	30% v/v 2-propanol
14	0.2M ammonium sulfate	0.1M Na cacodylate pH 6.5	30% w/v PEG 8000
15	0.2M magnesium acetate tetrahydrate	0.1M Na cacodylate pH 6.5	20% PEG 8000

Continued on next page

#	Salt	Buffer	Precipitant
16	0.2M magnesium acetate tetrahydrate	0.1M Na cacodylate pH 6.5	30% v/v 2-methyl-2,4-pentanediol
17	none	0.1M imidazole pH 6.5	1.0M sodium acetate trihydrate
18	0.2M sodium acetate trihydrate	0.1M Na cacodylate pH 6.5	30% w/v PEG 8000
19	0.2M zinc acetate dihydrate	0.1M Na cacodylate pH 6.5	18% w/v PEG 8000
20	0.2M calcium acetate hydrate	0.1M Na cacodylate pH 6.5	18% w/v PEG 8000
21	0.2M tri-sodium citrate dihydrate	0.1M Na HEPES pH 7.5	30% v/v 2-methyl-2,4-pentanediol
22	0.2M magnesium chloride hexahydrate	0.1M Na HEPES pH 7.5	30% v/v 2-propanol
23	0.2M calcium chloride dihydrate	0.1M Na HEPES pH 7.5	28% v/v PEG 400
24	0.2M magnesium chloride hexahydrate	0.1M Na HEPES pH 7.5	30% v/v PEG 400
25	0.2M tri-sodium citrate dihydrate	0.1M Na HEPES pH 7.5	20% v/v 2-propanol
26	none	0.1M Na HEPES pH 7.5	0.8M K, Na tartrate tetrahydrate
27	none	0.1M Na HEPES pH 7.5	1.5M lithium sulfate monohydrate
28	none	0.1M Na HEPES pH 7.5	0.8M Na dihydrogen phosphate / 0.8M K dihydrogen phosphate monohyd.
29	none	0.1M Na HEPES pH 7.5	1.4M tri-sodium citrate dihydrate
30	none	0.1M Na HEPES pH 7.5	2% v/v PEG 400, 2.0M amm. sulfate
31	none	0.1M Na HEPES pH 7.5	10% v/v 2-propanol, 20% w/v PEG 4000
32	none	0.1M Tris HCl pH 8.5	2.0M ammonium sulfate
33	0.2M magnesium chloride hexahydrate	0.1M Tris HCl pH 8.5	30% w/v PEG 4000
34	0.2M tri-sodium citrate dihydrate	0.1M Tris HCl pH 8.5	30% v/v PEG 400
35	0.2M lithium sulfate monohydrate	0.1M Tris HCl pH 8.5	30% w/v PEG 4000

Continued on next page

36	0.2M ammonium acetate	0.1M Tris HCl pH 8.5	30% v/v 2-propanol
37	0.2M sodium acetate trihydrate	0.1M Tris HCl pH 8.5	30% w/v PEG 4000
38	none	0.1M Tris HCl pH 8.5	8% w/v PEG 8000
39	none	0.1M Tris HCl pH 8.5	2.0M ammonium dihydrogen phosphate
40	none	none	0.4M K, Na tartrate tetrahydrate
41	none	none	0.4M ammonium dihydrogen phosphate
42	0.2M ammonium sulfate	none	30% w/v PEG 8000
43	0.2M ammonium sulfate	none	30% w/v PEG 4000
44	none	none	2.0M ammonium sulfate
45	none	none	4.0M sodium formate
46	0.05M potassium dihydrogen phosphate	none	20% w/v PEG 8000
47	none	none	30% w/v PEG 1500
48	none	none	0.2M magnesium formate
49	1.0M lithium sulfate monohydrate	none	2% w/v PEG 8000
50	0.5M lithium sulfate monohydrate	none	15% w/v PEG 8000

A5.9.

Screen name: **Structure Screen 2**
Supplier: **Molecular Dimension, Ltd.** (reprinted with permission)

	precipitant	buffer	salt
1	0.1 M sodium chloride	0.1 M bicine pH 9.0	30% w/v PEG monomethylether 550
2	none	0.1 M bicine pH 9.0	2.0 M magnesium chloride hexahydrate
3	2% w/v dioxane	0.1 M bicine pH 9.0	10% w/v PEG 20,000
4	0.2 M magnesium chloride hexahydrate	0.1 Tris pH 8.5	3.4 M 1,6-hexanediol
5	none	0.1 Tris pH 8.5	25% v/v *tert*-butanol
6	0.01 M nickel chloride hexahydrate	0.1 Tris pH 8.5	1.0 M lithium sulfate
7	1.5 M ammonium sulfate	0.1 Tris pH 8.5	12% v/v glycerol
8	0.2 M ammonium phosphate monobasic	0.1 Tris pH 8.5	50% v/v MPD
9	none	0.1 Tris pH 8.5	20% v/v ethanol
10	0.01 M nickel chloride hexahydrate	0.1 Tris pH 8.5	20% w/v PEG monomethylether 2000
11	0.5 M ammonium sulfate	0.1 M HEPES pH 7.5	30% v/v MPD
12	none	0.1 M HEPES pH 7.5	10% w/v PEG 6000, 5% v/v MPD
13	none	0.1 M HEPES pH 7.5	20% v/v Jeffamine M-600
14	0.1 M sodium chloride	0.1 M HEPES pH 7.5	1.6 M ammonium sulfate
15	none	0.1 M HEPES pH 7.5	2.0 M ammonium formate
16	0.05 M cadmium sulfate octahydrate	0.1 M HEPES pH 7.5	1.0 M sodium acetate

Continued on next page

17	none	0.1 M HEPES pH 7.5	70% v/v MPD
18	none	0.1 M HEPES pH 7.5	4.3 M sodium chloride
19	none	0.1 M HEPES pH 7.5	10% w/v PEG 8000 / 8% v/v ethylene glycol
20	none	0.1 M MES pH 6.5	1.6 M magnesium sulfate heptahydrate
21	0.1 M sodium phosphate monobasic / 0.1 M potassium phosphate monobasic	0.1 M MES pH 6.5	2.0 M sodium chloride
22	none	0.1 M MES pH 6.5	12% w/v PEG 20,000
23	1.6 M ammonium sulfate	0.1 M MES pH 6.5	10% v/v dioxane
24	0.05 M cesium chloride	0.1 M MES pH 6.5	30% v/v Jeffamine M-600
25	0.01 M cobalt chloride hexahydrate	0.1 M MES pH 6.5	1.8 M ammonium sulfate
26	0.2 M ammonium sulfate	0.1 M MES pH 6.5	30% w/v PEG monomethylether 5000
27	0.01 M zinc sulfate heptahydrate	0.1 M MES pH 6.5	25% v/v PEG monomethylether 550
28	none	0.1 M HEPES pH 7.5	20% w/v PEG 10,000
29	0.2 M K/Na tartrate	0.1 M sodium citrate pH 5.6	2.0 M ammonium sulfate
30	0.5 M ammonium sulfate	0.1 M sodium citrate pH 5.6	1.0 M lithium sulfate
31	0.5 M sodium chloride	0.1 M sodium citrate pH 5.6	4% v/v polyethyleneamine
32	none	0.1 M sodium citrate pH 5.6	35% v/v tert-butanol
33	0.01 M ferric chloride hexahydrate	0.1 M sodium citrate pH 5.6	10% v/v Jeffamine M-600
34	0.01 M manganese chloride tetrahydrate	0.1 M sodium citrate pH 5.6	2.5 M 1,6-hexanediol
35	none	0.1 M sodium acetate pH 4.6	2.0 M sodium chloride
36	0.2 M sodium chloride	0.1 M sodium acetate pH 4.6	30% v/v MPD

Continued on next page

Index